Houghton
Mifflin
Harcourt

GO MATH!
FLORIDA

FLORIDA

Houghton
Mifflin
Harcourt

Printed in the U.S.A.

ISBN 978-0-544-50077-8

1 2 3 4 5 6 7 8 9 10 0877 23 22 21 20 19 18 17 16 15

4500527063   ^ B C D E F G

Dear Students and Families,

Welcome to **Go Math!**, Kindergarten! In this exciting mathematics program, there are hands-on activities to do and real-world problems to solve. Best of all, you will write your ideas and answers right in your book. In **Go Math!**, writing and drawing on the pages helps you think deeply about what you are learning, and you will really understand math!

By the way, all of the pages in your **Go Math!** book are made using recycled paper. We wanted you to know that you can Go Green with **Go Math!**

Sincerely,

The Authors

Made in the United States
Text printed on 100% recycled paper

GO MATH!

Authors

Juli K. Dixon
Professor of Mathematics Education
University of Central Florida
Orlando, Florida

Matt Larson
Curriculum Specialist for Mathematics
Lincoln Public Schools
Lincoln, Nebraska

Miriam A. Leiva
Founding President, TODOS:
    Mathematics for All
Distinguished Professor
    of Mathematics Emerita
University of North Carolina Charlotte
Charlotte, North Carolina

Thomasenia Lott Adams
Professor of Mathematics Education
University of Florida
Gainesville, Florida

Houghton Mifflin Harcourt Publishing Company

Number and Operations

Representing, relating, and operating on whole numbers, initially with sets of objects

1 Represent, Count, Read, and Write Numbers 0 to 5 9

Domains Counting and Cardinality
Operations and Algebraic Thinking

2 Compare Numbers to 5 57

Domain Counting and Cardinality

DIGITAL PATH
Go online! Your math lessons are interactive. Use iTools, Animated Math Models, the Multimedia *e*Glossary, and more.

Math Story

Fall Festival!

Look for these:

REAL WORLD

H.O.T.
Higher Order Thinking
(see Teacher Edition)

Use every day
For Standards Practice.

© Houghton Mifflin Harcourt Publishing Company

v

Look for these:

REAL WORLD

H.O.T.
Higher Order Thinking
(see Teacher Edition)

GO MATH! FLORIDA

Use every day
For Standards Practice.

© Houghton Mifflin Harcourt Publishing Company

Look for these:

REAL WORLD

H.O.T.
Higher Order Thinking
(see Teacher Edition)

GO MATH!
FLORIDA

Use every day
For Standards Practice.

Geometry and Positions

Describing shapes and space

DIGITAL PATH
Go online! Your math lessons are interactive. Use iTools, Animated Math Models, the Multimedia *eGlossary*, and more.

Math Story

School Fun

Look for these:

REAL WORLD

H.O.T.
Higher Order Thinking
(see Teacher Edition)

GO MATH! FLORIDA

Use every day
For Standards Practice.

10 Identify and Describe Three-Dimensional Shapes 409

Domain Geometry

Measurement and Data

Representing, relating, and operating on whole numbers, initially with sets of objects

Fall Festival!

written by Alison Juliano

Representing, relating, and operating on whole numbers, initially with sets of objects

Fall is here! What do you see?

One big apple tree.

What season is this?

Fall is here! What do you see?

Two pumpkins for you and me.

Science

What do you know about fall?

three **3**

Fall is here! What do you see?

Bales of hay—1, 2, 3!

Science

What do people wear in fall?

Fall is here! What do you see?

Four leaves falling from a tree.

Science

What changes in fall?

Fall is here! What do you see?

Five stalks of corn. Do you see me?

Science

How is fall different from the other seasons?

Write About the Story

Vocabulary Review

one	four
two	five
three	

DIRECTIONS Look at the picture of the fall scene. Using the numbers you have learned, draw a story about fall. Invite a friend to count the objects in your story.

Count How Many

 1. 　1　2　3　④　5

 2. 　1　2　3　4　⑤

 3. 　1　2　③　4　5

 4. 　①　2　3　4　5

 5. 　1　②　3　4　5

DIRECTIONS 1–5. Look at the picture. Count how many.
Circle the number.

8 eight

Represent, Count, Read, and Write Numbers 0 to 5

Curious About Math with

Curious George

Navel oranges have no seeds.

- How many seeds do you see?

Name _____

Show What You Know ✓

Explore Numbers

 1

Match Numbers to Sets

| 1 | 2 | 3 | 4 | 5 |

DIRECTIONS 1. Circle all of the sets of three oranges. 2. Draw a line to match the number to the set.

© Houghton Mifflin Harcourt Publishing Company

FAMILY NOTE: This page checks your child's understanding of important skills needed for success in Chapter 1.

GO Online   Assessment Options
Soar to Success Math

Name _____

Vocabulary Builder

match

set

DIRECTIONS Draw a line to match a set of chicks to a set of flowers.

• eStudent Edition
• Multimedia eGlossary

Game Bus Stop

DIRECTIONS Each player rolls the number cube. The first player to roll a 1 moves to the bus stop marked 1. Continue playing until each player has rolled the numbers in sequence and stopped at each bus stop. The first player to reach 5 wins the game.

MATERIALS game marker for each player, number cube (0–5).

Compare Numbers to 5

Curious About Math with **Curious George**

Butterflies have taste buds in their feet so they stand on their food to taste it!

- Are there more butterflies or more flowers in this picture?

Name _____

One-to-One Correspondence

 1

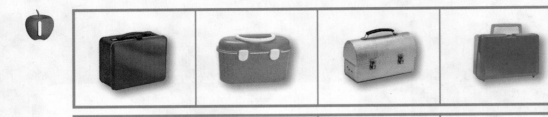

Model Numbers 0 to 5

2

Write Numbers 0 to 5

3  _____

4  _____

DIRECTIONS **1.** Draw one apple for each lunch box. **2.** Place counters in the five frame to model the number. Draw the counters. Trace the number. **3–4.** Count and tell how many. Write the number.

FAMILY NOTE: This page checks your child's understanding of important skills needed for success in Chapter 2.

GO Online
Assessment Options
Soar to Success Math

Name _____

Vocabulary Builder

four

three

three

one

two

five

DIRECTIONS Circle the sets with the same number of animals. Count and tell how many trees. Draw a line below the word for the number of trees.

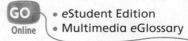
• eStudent Edition
• Multimedia eGlossary

Chapter 2

Game

Counting to Blastoff

Player 1

5	4	3	2	1	0

Player 2

5	4	3	2	1	0

DIRECTIONS Each player tosses the number cube and finds that number on his or her board. The player covers the number with a counter. Players take turns in this way until they have covered all of the numbers on the board. Then they are ready for blast off.

MATERIALS 6 counters for each player, number cube (0–5)

Name _____

Same Number

Essential Question How can you use matching and counting to compare sets with the same number of objects?

Listen and Draw REAL WORLD

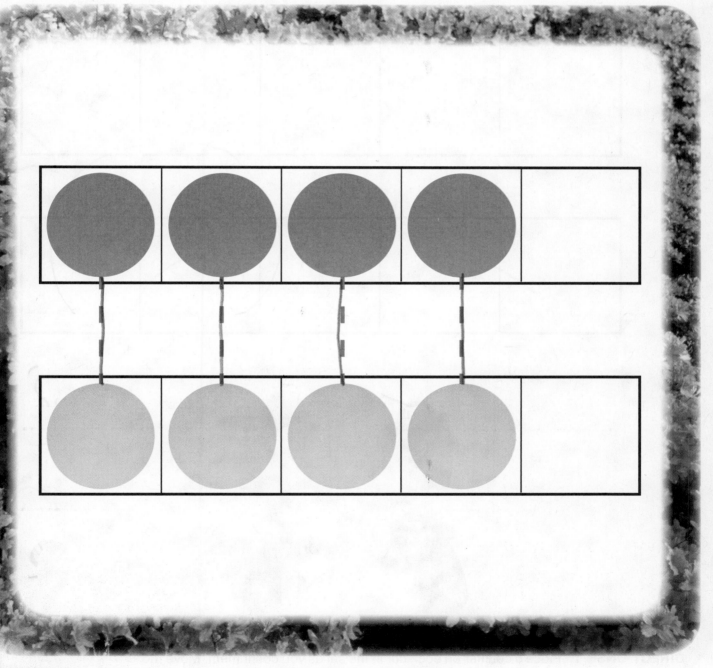

DIRECTIONS Place counters as shown. Trace the lines to match each counter in the top five frame to a counter below it in the bottom five frame. Count how many in each set. Tell a friend about the number of counters in each set.

Share and Show

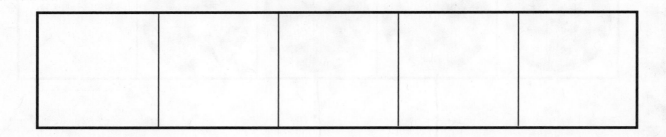

DIRECTIONS 1. Place a counter on each car in the set as you count them. Move the counters to the five frame below the cars. Draw the counters. Place a counter on each finger puppet in the set as you count them. Move the counters to the five frame above the puppets. Draw those counters. Is the number of objects in one set greater than, less than, or the same as the number of objects in the other set? Draw a line to match a counter in each set.

62 sixty-two

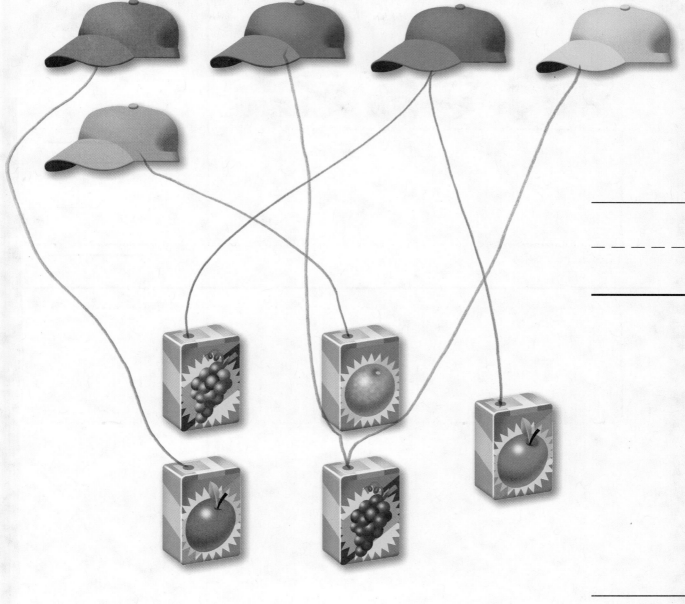

DIRECTIONS 2. Compare the sets of objects. Is the number of hats greater than, less than, or the same as the number of juice boxes? Count how many hats. Write the number. Count how many juice boxes. Write the number. Tell a friend what you know about the number of objects in each set.

PROBLEM SOLVING REAL WORLD

 1

_____

2 _____

_____
_____

B _____

_____

2

DIRECTIONS **1.** Count how many buses. Write the number. Draw to show a set of counters that has the same number as the set of buses. Write the number. Draw a line to match the objects in each set. **2.** Draw two sets that have the same number of objects shown in different ways. Tell a friend about your drawing.

HOME ACTIVITY · Show your child two sets that have the same number of up to five objects. Have him or her identify whether the number of objects in one set is greater than, less than, or has the same number of objects as the other set.

FOR MORE PRACTICE:
Standards Practice Book, pp. P27–P28

Name _____

Greater Than

Essential Question How can you compare sets when the number of objects in one set is greater than the number of objects in the other set?

Listen and Draw

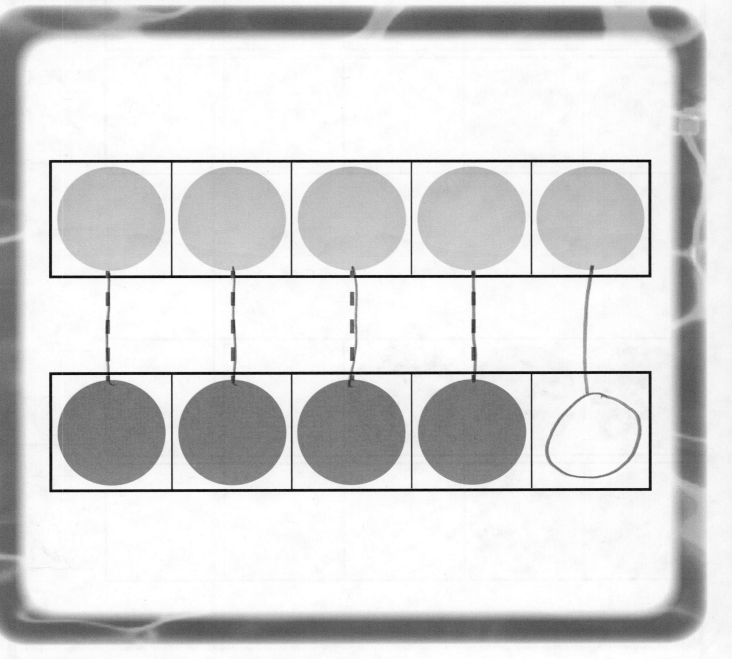

DIRECTIONS Place counters as shown. Trace the lines to match a counter in the top five frame to a counter below it in the bottom five frame. Count how many in each set. Tell a friend which set has a number of objects greater than the other set.

Chapter 2 • Lesson 2

Share and Show

 1

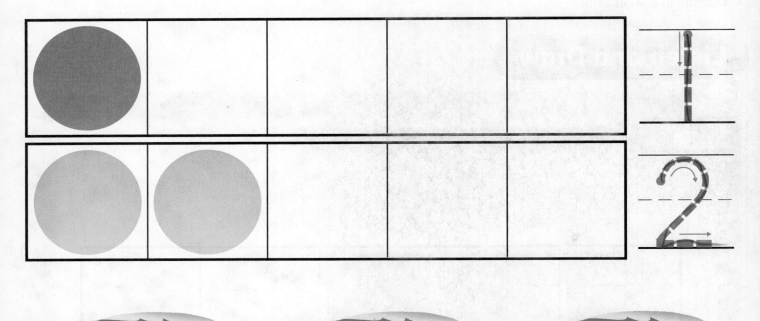

 2

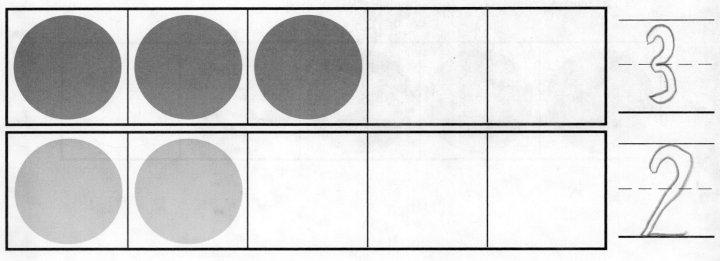

DIRECTIONS 1. Place counters as shown. Count and tell how many in each set. Trace the numbers. Compare the sets by matching. Circle the number that is greater. 2. Place counters as shown. Count and tell how many in each set. Write the numbers. Compare the sets by matching. Circle the number that is greater.

66 sixty-six

Name ______________________________

3

3 4

4

2 1

DIRECTIONS 3–4. Place counters as shown. Count and tell how many in each set. Write the numbers. Compare the numbers. Circle the number that is greater.

PROBLEM SOLVING REAL WORLD

DIRECTIONS Brianna has a bag with three apples in it. Her friend has a bag with a set of apples that is one larger. Draw the bags. Write the numbers on the bags to show how many apples. Tell a friend what you know about the numbers.

HOME ACTIVITY · Show your child a set of up to four objects. Have him or her show a set with a number of objects greater than your set.

FOR MORE PRACTICE:
Standards Practice Book, pp. P29–P30

Name _____

Less Than

Essential Question How can you compare sets when the number of objects in one set is less than the number of objects in the other set?

Listen and Draw

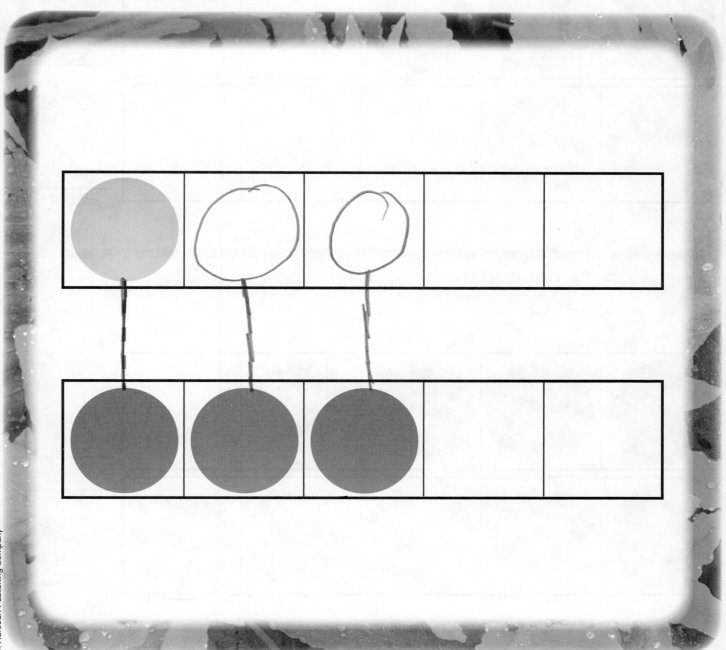

DIRECTIONS Place counters as shown. Trace the line to match a counter in the top five frame to a counter below it in the bottom five frame. Count how many in each set. Tell a friend which set has a number of objects less than the other set.

Chapter 2 • Lesson 3

Share and Show

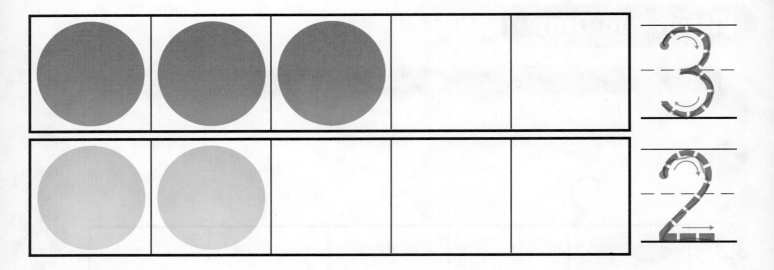

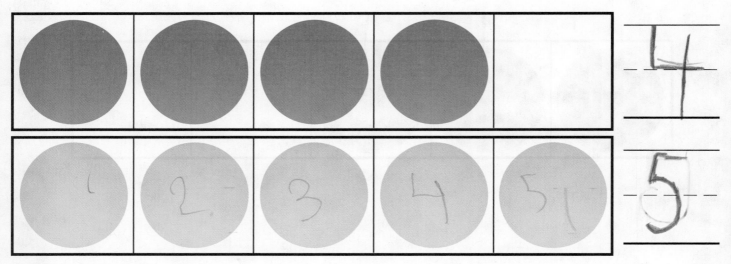

DIRECTIONS **1.** Place counters as shown. Count and tell how many in each set. Trace the numbers. Compare the sets by matching. Circle the number that is less. **2.** Count and tell how many in each set. Write the numbers. Compare the sets by matching. Circle the number that is less.

70 seventy

 3

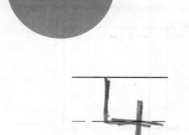

4

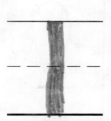

DIRECTIONS   3–4. Count and tell how many in each set. Write the numbers. Compare the numbers. Circle the number that is less.

HOME ACTIVITY · Show your child a set of two to five objects. Have him or her show a set of objects that has a number of objects less than your set.

FOR MORE PRACTICE:
Standards Practice Book, pp. P31–P32

Chapter 2 · Lesson 3

Mid-Chapter Checkpoint

Concepts and Skills

1

_____

_ _ _ _ _ _ _ _

_____

_ _ _ _ _ _ _ _

_____

2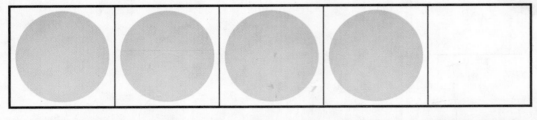

_ _ _ _ _ _ _ _

_____

_ _ _ _ _ _ _ _

_____

3

⚪   ⚪   ⚪   ⚪

DIRECTIONS **1.** Place a counter below each object to show the same number of objects. Draw and color each counter. Write how many objects in each row. **2.** Place counters as shown. Count and tell how many in each set. Write the numbers. Compare the sets by matching. Circle the number that is greater. **3.** Count the fish in the bowl at the beginning of the row. Mark under the bowl that has a number of fish less than the bowl at the beginning of the row.

Name _____

Problem Solving • Compare by Matching Sets to 5

Essential Question How can you make a model to solve problems using a matching strategy?

🔑 Unlock the Problem REAL WORLD

DIRECTIONS These are Brandon's toy cars. How many toy cars does Brandon have? Jay has a number of toy cars that is less than the number of toy cars Brandon has. Use cubes to show how many toy cars Jay might have. Draw the cubes. Use matching to compare the sets.

Chapter 2 • Lesson 4

seventy-three **73**

Try Another Problem

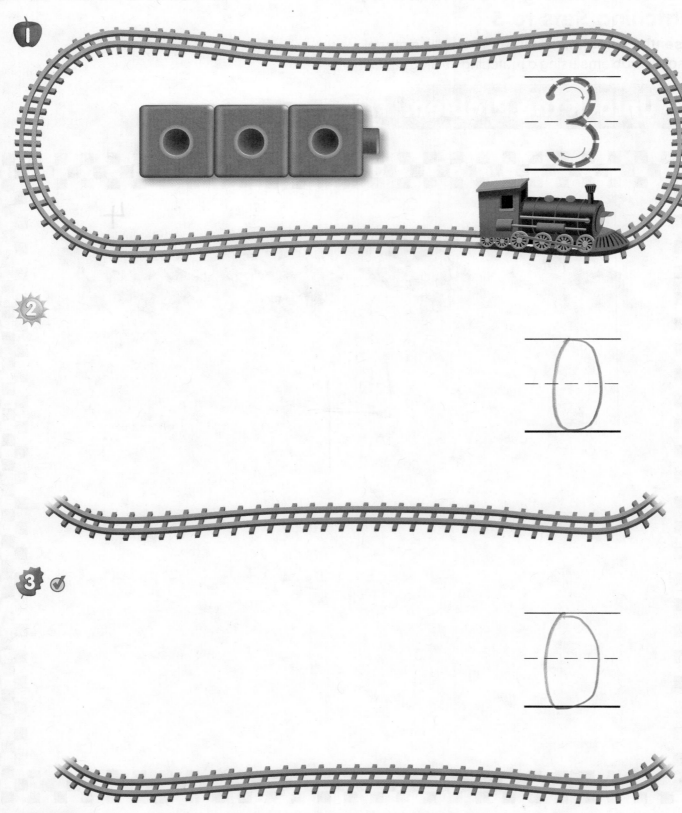

1

2

3 ✓

DIRECTIONS 1. How many cubes? Trace the number. 2–3. Model a cube train that has a number of cubes greater than 3. Draw the cube train. Write how many. Compare the cube train by matching with the model at the top of the page. Tell a friend about the cube trains.

Name _____

Share and Show

4

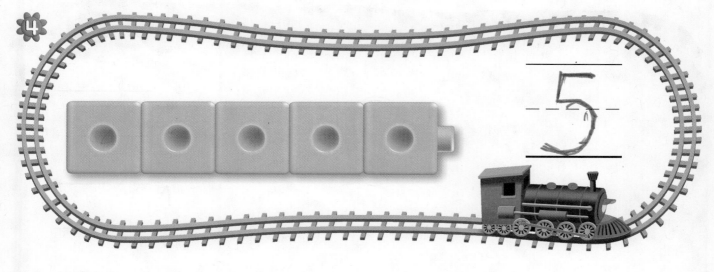

5

5

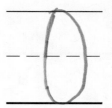

6

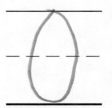

DIRECTIONS **4.** How many cubes? Write the number. **5–6.** Model a cube train that has a number of cubes less than 5. Draw the cube train. Write how many. Compare the cube train by matching with the model at the top of the page. Tell a friend about the cube trains.

Chapter 2 • Lesson 4

seventy-five **75**

On Your Own REAL WORLD

1

_____

- - - - - - - - - - - -

_____

_____

- - - - - - - - - - - -

_____

2

_____

- - - - - - - - - - - -

_____

_____

- - - - - - - - - - - -

_____

DIRECTIONS **1.** Kendall has a set of three pencils. Her friend has a set with the same number of pencils. Draw to show the sets of pencils. Compare the sets by matching. Write how many in each set. **2.** Draw to show what you know about matching to compare two sets of objects. Write how many in each set.

HOME ACTIVITY · Show your child two sets with a different number of objects in each set. Have him or her use matching to compare the sets.

FOR MORE PRACTICE:
Standards Practice Book, pp. P33–P34

Name _____

Compare by Counting Sets to 5

Essential Question How can you use a counting strategy to compare sets of objects?

Listen and Draw REAL WORLD

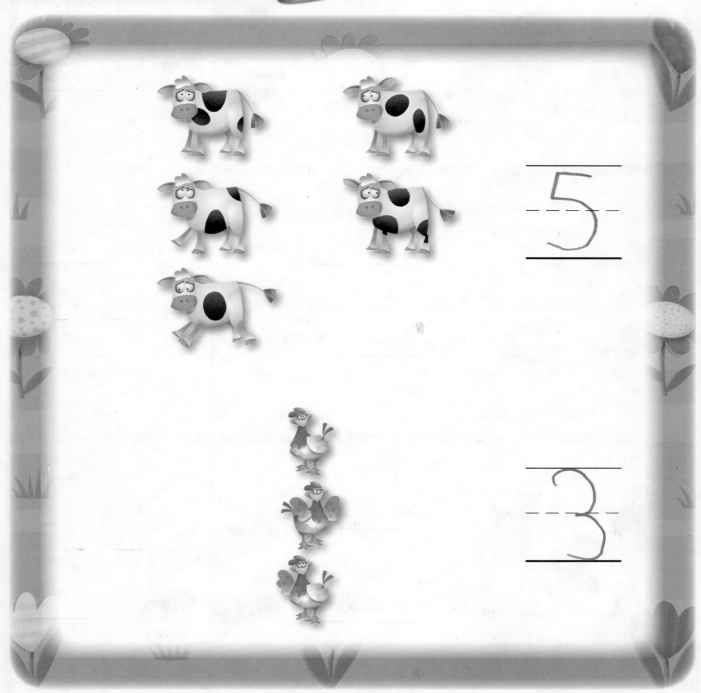

DIRECTIONS Look at the sets of objects. Count how many objects in each set. Write the numbers. Compare the numbers and tell a friend which number is greater and which number is less.

Share and Show

1  _____ 2 _____  _____ 3 _____

2  _____ 3 _____  _____ 1 _____

3 ✓  _____ 5 _____  _____ 4 _____

DIRECTIONS 1–3. Count how many objects in each set. Write the numbers. Compare the numbers. Circle the number that is greater.

4

 1

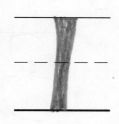 2

5

 5

 4

6

 3

 5

DIRECTIONS 4–6. Count how many objects in each set.
Write the numbers. Compare the numbers. Circle the number
that is less.

PROBLEM SOLVING REAL WORLD

1    3

   3

2

   5

   2

DIRECTIONS   **1.** Count how many objects in each set. Write the numbers. Compare the numbers. Tell a friend what you know about the sets.   **2.** Draw to show what you know about counting to compare two sets of objects. Write how many in each set.

HOME ACTIVITY · Draw a domino block with up to three dots on one end. Ask your child to draw on the other end a set of dots that has a number of dots greater than the set you drew.

FOR MORE PRACTICE:
Standards Practice Book, pp. P35–P36

Name _____

Vocabulary

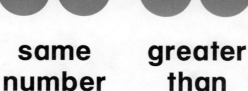

same number   **greater than**   **less than**

Concepts and Skills

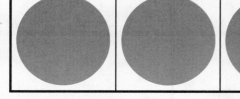   4

   3

   5      4

DIRECTIONS   **1.** Draw a line from the words to the sets of counters to show the same number, greater than, and less than the number of hats at the beginning of the row. (pp. 61, 65, 69)   **2.** Count and tell how many in each set. Write the numbers. Compare the sets by matching. Circle the number that is greater.   **3.** Count and tell how many in each set. Write the numbers. Compare the numbers. Circle the number that is less.

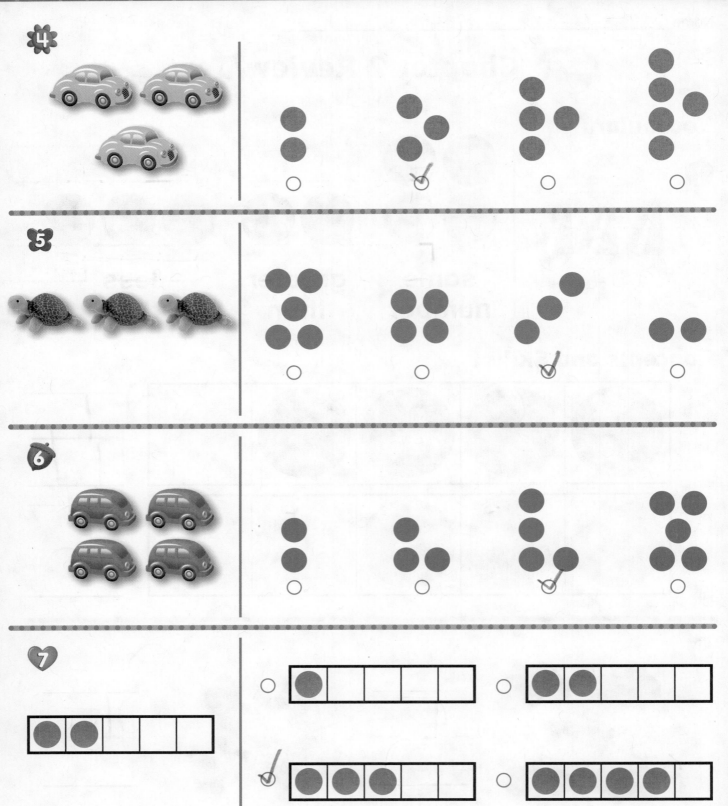

DIRECTIONS **4.** Mark under the set that has the same number of counters as the number of cars at the beginning of the row. **5.** Mark under the set that has a number of counters less than the number of turtles at the beginning of the row. **6.** Mark under the set that has a number of counters greater than the number of vans at the beginning of the row. **7.** Mark beside the set that shows a number of counters less than the number of counters at the beginning of the row.

82 eighty-two

8

○ ◉◉▢▢ ○ ◉◉◉▢

◉◉◉◉▢

✓ ◉◉◉◉▢ ○ ◉◉◉◉◉

9

1 ○ 2 ○ 3 ○ 4 ✓

10

2 ○ 3 ○ 4 ✓ 5 ○

DIRECTIONS **8.** Mark beside the set that shows a number of counters greater than the number of counters at the beginning of the row. **9.** Mark under the number that is greater than 3. **10.** Mark under the number that is less than 3.

Performance Task

_____

- - - - - - - -

_____

_____

- - - - - - - -

_____

PERFORMANCE TASK   This task will assess the child's understanding of comparing and ordering numbers to 5.

Represent, Count, Read, and Write Numbers 6 to 9

Curious About Math with Curious George

Rides are popular at fairs.

- What can you tell me about this ride?

Name _____

 Show What You Know ✓

Explore Numbers to 5

 1

 2

Compare Numbers to 5

 3

3

 2

Write Numbers to 5

 4 0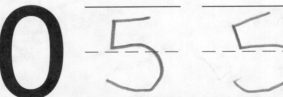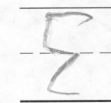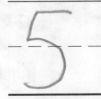

DIRECTIONS **1.** Circle the dot cards that show 3. **2.** Circle the dot cards that show 5. **3.** Write the number of cubes in each set. Circle the greater number. **4.** Write the numbers 1 to 5 in order.

FAMILY NOTE: This page checks your child's understanding of important skills needed for success in Chapter 3.

GO Online Assessment Options
Soar to Success Math

Name _____

Vocabulary Builder

same number

more

fewer

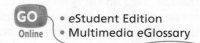

DIRECTIONS Point to sets of objects as you count. Circle two sets that have the same number of objects. Tell what you know about sets that have more objects or fewer objects than other sets on this page.

GO Online
• eStudent Edition
• Multimedia *eGlossary*

Game
Number Line Up

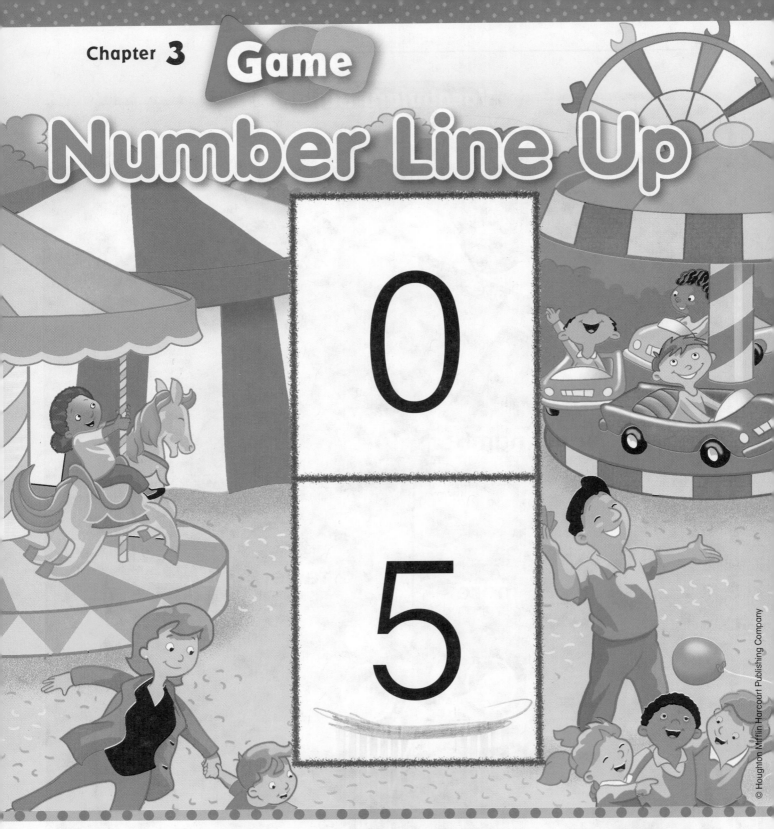

DIRECTIONS Play with a partner. Place numeral cards as shown on the board. Shuffle the remaining cards and place them face down in a stack. Players take turns picking one card from the stack. They place the card to the right to form a number sequence without skipping any numbers. The number sequence can be forward from 0 or backward from 5. If a player picks a card that is not next in either number sequence, the card is returned to the bottom of the stack. The first player to complete a number sequence wins the game.

MATERIALS 2 sets of numeral cards 0–5

Name _____

Model and Count 6

Essential Question How can you show and count
6 objects?

Listen and Draw   REAL WORLD

1	2	3	4	5
6				

DIRECTIONS Place a counter on each ticket in the set as you
count them. Move the counters to the ten frame. Draw the counters.

Chapter 3 • Lesson 1   eighty-nine **89**

Share and Show

six

DIRECTIONS 1. Place a counter on each car in the set as you count them. Move the counters to the parking lot. Draw the counters. Say the number as you trace it.

②

6
six

___ ___ ___ ___

and

and

and

and

___ ___ ___ ___

DIRECTIONS 2. Trace the number 6. Use two-color counters to model the different ways to make 6. Write to show some pairs of numbers that make 6.

Chapter 3 • Lesson 1

ninety-one **91**

PROBLEM SOLVING REAL WORLD

1.

2.

DIRECTIONS 1. Count the objects in each set. Which sets show six objects? Circle those sets. 2. Draw to show what you know about a set of six objects. Tell about your drawing.

HOME ACTIVITY • Ask your child to show a set of five objects. Have him or her show one more object and tell how many objects are in the set.

FOR MORE PRACTICE:
Standards Practice Book, pp. P41–P42

Name _____

Read and Write to 6

Essential Question How can you read and write 6 with words and numbers?

Listen and Draw REAL WORLD

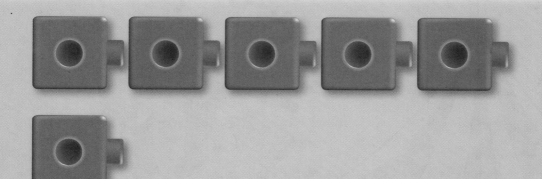

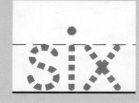

DIRECTIONS Count and tell how many cubes. Trace the numbers. Count and tell how many hats. Trace the word.

Chapter 3 • Lesson 2

Share and Show

1

DIRECTIONS **1.** Look at the picture. Circle the sets of six objects, or people.

Name _____

2

6
six

6 6 6 6 6

3 ☑

_____

_ _ _ _ _ _ _

_____

4 ☑

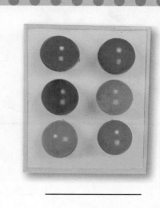

_____

_ _ _ _ _ _ _

_____

5

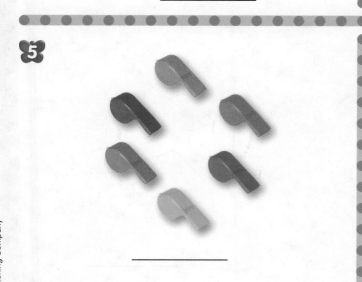

_____

_ _ _ _ _ _ _

_____

6

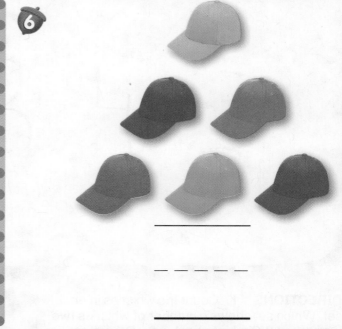

_____

_ _ _ _ _ _ _

_____

DIRECTIONS **2.** Say the number. Trace the numbers.
3–6. Count and tell how many. Write the number.

PROBLEM SOLVING  REAL WORLD

1

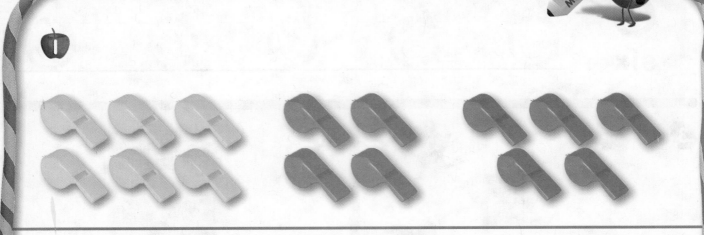

2

_____

- - - - -

_____

DIRECTIONS **1.** Count the whistles in each set. Which set shows a number of whistles two less than 6? Circle that set.   **2.** Draw a set of objects that has a number of objects one greater than 5. Tell about your drawing. Write how many objects.

HOME ACTIVITY • Show six objects. Have your child point to each object as he or she counts it. Then have him or her write the number on paper to show how many.

96 ninety-six

FOR MORE PRACTICE:
Standards Practice Book, pp. P43–P44

Name _____

Model and Count 7

Essential Question How can you show and count 7 objects?

Listen and Draw

DIRECTIONS Model 6 objects. Show one more object. How many are there? Tell a friend how you know. Draw the objects.

© Houghton Mifflin Harcourt Publishing Company

Share and Show

1

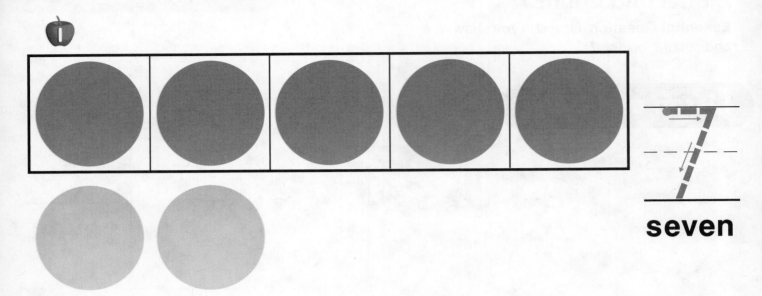

7

seven

2

5 and ____ more

3 ☑

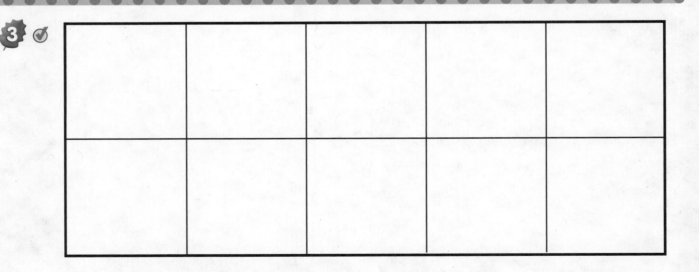

DIRECTIONS **I.** Place counters as shown. Count and tell how many counters. Trace the number. **2.** How many more than 5 is 7? Write the number. **3.** Place counters in the ten frame to model seven. Tell a friend what you know about the number 7.

❀ 4

7

seven

_____   **and**   _____

_____   **and**   _____

_____   **and**   _____

_____   **and**   _____

DIRECTIONS   4. Use two-color counters to model the different ways to make 7. Write to show some pairs of numbers that make 7.

PROBLEM SOLVING REAL WORLD

DIRECTIONS 1. Count the objects in each set. Which sets show seven objects? Circle those sets. **2.** Draw to show what you know about the number 7. Tell a friend about your drawing.

HOME ACTIVITY • Ask your child to show a set of six objects. Have him or her show one more object and tell how many objects are in the set.

FOR MORE PRACTICE:
Standards Practice Book, pp. P45–P46

Name _____

Read and Write to 7

Essential Question How can you read and write 7 with words and numbers?

Listen and Draw REAL WORLD

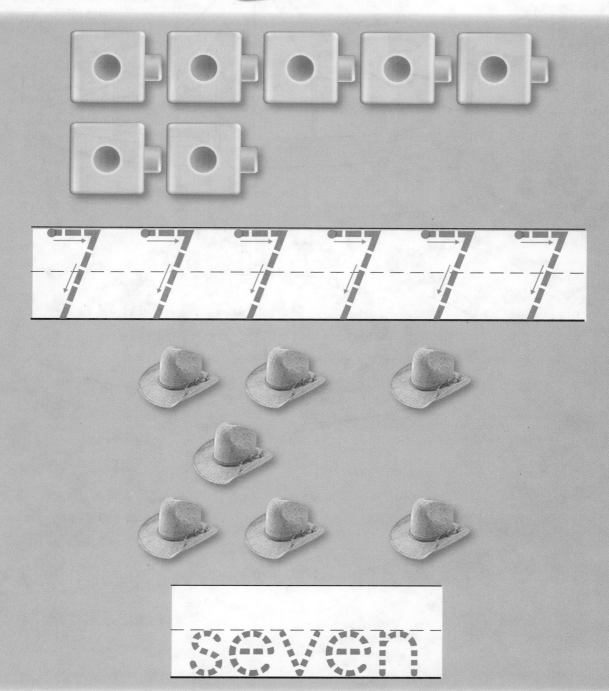

DIRECTIONS Count and tell how many cubes. Trace the numbers. Count and tell how many hats. Trace the word.

Chapter 3 • Lesson 4

© Houghton Mifflin Harcourt Publishing Company

Share and Show

DIRECTIONS I. Look at the picture. Circle the sets of seven objects.

Name _____

7
seven

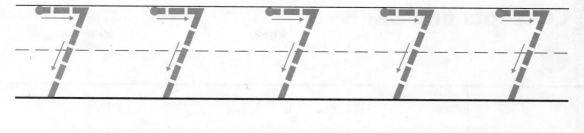

 _____

_ _ _ _ _ _ _ _

_____

 _____

_ _ _ _ _ _ _ _

_____

 _____

_ _ _ _ _ _ _ _

_____

 _____

_ _ _ _ _ _ _ _

_____

DIRECTIONS 2. Say the number. Trace the numbers. **3–6.** Count and tell how many. Write the number.

HOME ACTIVITY • Show your child seven objects. Have him or her point to each object as he or she counts it. Then have him or her write the number on paper to show how many objects.

Chapter 3 • Lesson 4

FOR MORE PRACTICE:
Standards Practice Book, pp. P47–P48

Concepts and Skills

1

2

_____

- - - - - - -

_____

3

_____

- - - - - - -

_____

 4

6   7   8   9
○   ○   ○   ○

DIRECTIONS **1.** Use counters to model the number 7. Draw the counters. Write the number. **2–3.** Count and tell how many. Write the number. **4.** Mark under the number that shows how many whistles are in the set at the beginning of the row.

Name _____

Model and Count 8

Essential Question How can you show and count 8 objects?

Listen and Draw

DIRECTIONS Model 7 objects. Show one more object. How many are there? Tell a friend how you know. Draw the objects.

Share and Show

1

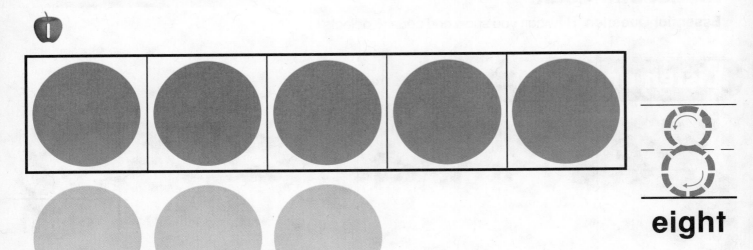

eight

2

5 and ____ more

---

3 ✓

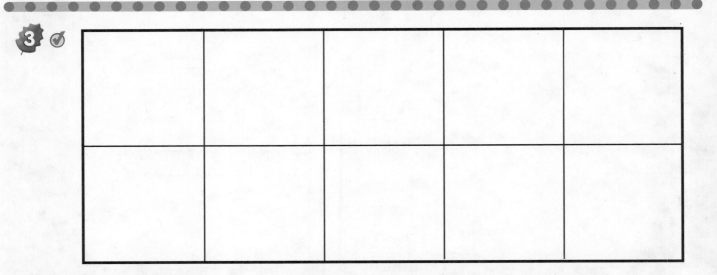

DIRECTIONS 1. Place counters as shown. Count and tell how many counters. Trace the number. 2. How many more than 5 is 8? Write the number. 3. Place counters in the ten frame to model eight. Tell a friend what you know about the number 8.

Name _____

4

8
eight

_____ ⬤ **and** _____ ⬤

_____ ⬤ **and** _____ ⬤

_____ ⬤ **and** _____ ⬤

_____ ⬤ **and** _____ ⬤

DIRECTIONS **4.** Use two-color counters to model the different ways to make 8. Write to show some pairs of numbers that make 8.

PROBLEM SOLVING REAL WORLD

1

2

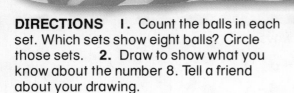

DIRECTIONS 1. Count the balls in each set. Which sets show eight balls? Circle those sets. **2.** Draw to show what you know about the number 8. Tell a friend about your drawing.

HOME ACTIVITY • Ask your child to show a set of seven objects. Have him or her show one more object and tell how many.

108 one hundred eight

FOR MORE PRACTICE:
Standards Practice Book, pp. P49–P50

Name _____

Read and Write to 8

Essential Question How can you read and write 8 with words and numbers?

Listen and Draw REAL WORLD

DIRECTIONS Count and tell how many cubes. Trace the numbers. Count and tell how many balls. Trace the word.

Share and Show

DIRECTIONS 1. Look at the picture. Circle the sets of eight objects.

Name _____

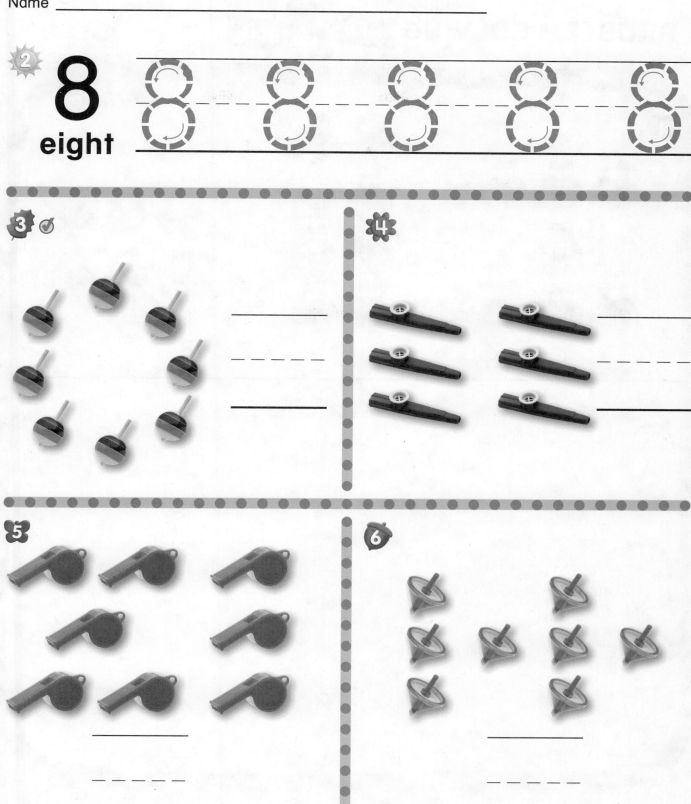

8
eight

DIRECTIONS **2.** Say the number. Trace the numbers.
3–6. Count and tell how many. Write the number.

Chapter 3 · Lesson 6

one hundred eleven **111**

© Houghton Mifflin Harcourt Publishing Company

PROBLEM SOLVING REAL WORLD

1

2

_____

_ _ _ _ _ _ _ _

_____

DIRECTIONS **1.** Count the objects in each set. Which set has a number of objects two greater than 6? Circle that set. **2.** Robbie won ten prizes at the fair. Marissa won a number of prizes two less than Robbie. Draw to show Marissa's prizes. Write how many.

HOME ACTIVITY • Show eight objects. Have your child point to each object as he or she counts it. Then have him or her write the number on paper to show how many objects.

112 one hundred twelve

Name _____

Model and Count 9

Essential Question How can you show and count 9 objects?

Listen and Draw

DIRECTIONS Model 8 objects. Show one more object. How many are there? Tell a friend how you know. Draw the objects.

Share and Show

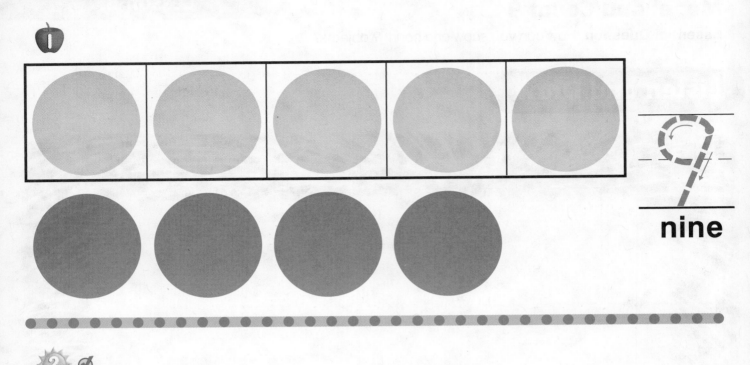

nine

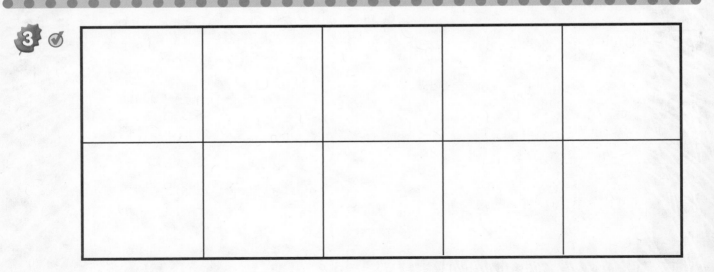

DIRECTIONS
1. Place counters as shown. Count and tell how many counters. Trace the number. **2.** How many more than 5 is 9? Write the number. **3.** Place counters in the ten frame to model nine. Tell a friend what you know about the number 9.

Name _____

⚜ 4

9
nine

_____ _____

_____ and _____

_____ and _____

_____ and _____

_____ and _____

DIRECTIONS 4. Use two-color counters to model the different ways to make 9. Write to show some pairs of numbers that make 9.

Chapter 3 • Lesson 7   one hundred fifteen **115**

PROBLEM SOLVING REAL WORLD

1.

2.

DIRECTIONS 1. Count the objects in each set. Which sets show nine objects? Circle those sets. **2.** Draw to show what you know about the number 9. Tell a friend about your drawing.

HOME ACTIVITY • Ask your child to show a set of eight objects. Have him or her show one more object and tell how many.

FOR MORE PRACTICE:
Standards Practice Book, pp. P53–P54

Name _____

Read and Write to 9

Essential Question How can you read and write 9 with words and numbers?

Listen and Draw  REAL WORLD

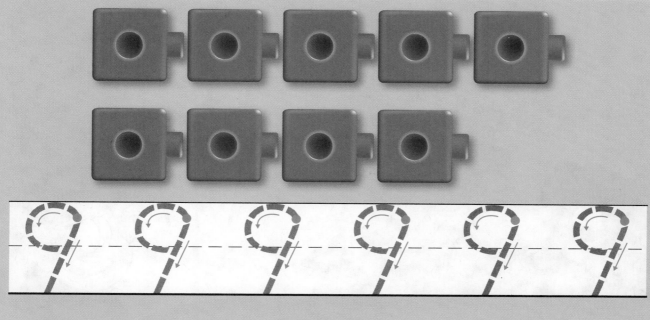

DIRECTIONS Count and tell how many cubes. Trace the numbers. Count and tell how many ducks. Trace the word.

Chapter 3 · Lesson 8

one hundred seventeen **117**

Share and Show

DIRECTIONS 1. Look at the picture. Circle the sets of nine objects.

Name _____

9
nine

9 9 9 9 9

---

3 ✓

_____

_ _ _ _ _ _ _

_____

4

_____

_ _ _ _ _ _ _

_____

5

_____

_ _ _ _ _ _ _

_____

6

_____

_ _ _ _ _ _ _

_____

DIRECTIONS **2.** Say the number. Trace the numbers.
3–6. Count and tell how many. Write the number.

PROBLEM SOLVING REAL WORLD

1

2

_____

_ _ _ _ _

_____

DIRECTIONS **1.** Which set has a number of objects one less than 10? Circle that set. **2.** Draw a set that has a number of objects two greater than 7. Write how many.

HOME ACTIVITY • Ask your child to find something in your home that has the number 9 on it, such as a clock or a phone.

FOR MORE PRACTICE:
Standards Practice Book, pp. P55–P56

Name _____

Problem Solving • Numbers to 9

Essential Question How can you solve problems using the strategy *draw a picture*?

🔑 Unlock the Problem REAL WORLD

DIRECTIONS There are seven flags on the red tent. Trace the flags. The blue tent has a number of flags one greater than the red tent. How many flags are on the blue tent? Draw the flags. Tell a friend about your drawing.

Chapter 3 • Lesson 9

Try Another Problem

DIRECTIONS **1.** Bianca buys five cupcakes. Leigh buys a number of cupcakes two greater than 5. Draw the cupcakes. Write the numbers. **2.** Donna wins nine tokens. Jackie wins a number of tokens two less than 9. Draw the tokens. Write the numbers.

Name _____

Share and Show

3 ✓

4

DIRECTIONS **3.** Gary has eight tickets. Four of the tickets are red. The rest are blue. How many are blue? Draw the tickets. Write the number beside each set of tickets. **4.** Ann has seven balloons. Molly has a set of balloons less than seven. How many balloons does Molly have? Draw the balloons. Write the number beside each set of balloons.

On Your Own

1

_____

- - - - - - - - -

_____

_____

- - - - - - - - -

_____

2

DIRECTIONS **1.** There are six seats on a teacup ride. The number of seats on a train ride is two less than 8. How many seats on the train ride? Draw the seats. Write the numbers. **2.** Pick two numbers between 0 and 9. Draw to show what you know about those numbers.

HOME ACTIVITY • Have your child say two different numbers from 0–9 and tell what he or she knows about them.

FOR MORE PRACTICE:
Standards Practice Book, pp. P57–P58

Chapter 3 Review/Test

Vocabulary

 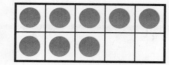

nine   **six**   **eight**

Concepts and Skills

②

③

5 and ____ more

© Houghton Mifflin Harcourt Publishing Company

DIRECTIONS   1. Draw a line to match the counters in the ten frame to the number word. (pp. 93, 109, 117)   **2.** Place counters in the ten frame to model seven. Draw the counters. Write the number.   **3.** How many more than 5 is 7? Write the number.

Chapter 3   one hundred twenty-five   **125**

4

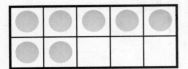

2 ○

3 ○

5 ○

7 ○

5

6 ○

8 ○

9 ○

10 ○

6

8

○   ○   ○   ○

7

9

○   ○   ○   ○

DIRECTIONS **4–5.** Mark under the number that shows how many. **6–7.** Mark under the set that models the number at the beginning of the row.

Name _____

8 🐟

○   ○   ○   ○

9 🐚

○   ○   ○   ○

10 🔟

$$7$$

○   ○   ○   ○

11 🐢

●	●	●	●	●
●	●	●		

6   7   8   9
○   ○   ○   ○

12 🔺

●	●	●	●	●
●	●	●	●	

9   8   7   6
○   ○   ○   ○

DIRECTIONS 8. Mark under the set that models a way to make 6.   **9.** Mark under the set that models a way to make 9.   **10.** Mark under the set that models the number at the beginning of the row.   **11–12.** Mark under the number that shows how many.

Performance Task

_____

- - - - - - -

_____

_____

- - - - - - -

_____

_____

- - - - - - -

_____

_____

- - - - - - -

_____

PERFORMANCE TASK This task will assess the child's understanding of representing numbers 6 to 9.

Represent and Compare Numbers to 10

Curious About Math with Curious George

Apple trees grow from a small seed.

• About how many seeds are in an apple?

Name _____

Show What You Know ✓

Draw Objects to 9

 1

9

 ②

7

Write Numbers to 9

 **3**

_____

- - - - - - - - - - -

_____

4

_____

- - - - - - - - - - -

_____

5

_____

- - - - - - - - - - -

_____

6

_____

- - - - - - - - - - -

_____

DIRECTIONS   1. Draw 9 flowers.
2. Draw 7 flowers.   **3–6.** Count and tell how
many. Write the number.

FAMILY NOTE: This page checks your child's
understanding of important skills needed for
success in Chapter 4.

GO Online   Assessment Options
Soar to Success Math

Vocabulary Builder

greater

less

same number

DIRECTIONS Circle the words that describe the number of carrots and the number of celery sticks. Use *greater* and *less* to describe the number of trees and the number of bushes.

GO Online
• eStudent Edition
• Multimedia eGlossary

Game

Spin and Count!

START

END

Spinner: 6, 7, 8, 9, 10, 11, 12

DIRECTIONS Play with a partner. Place game markers on START. Use a pencil and a paper clip to spin for a number. Take turns spinning. Each player moves his or her marker to the next space that has the same number of objects as the number on the spinner. The first player to reach END wins.

MATERIALS two game markers, pencil, paper clip

Name _____

Model and Count 10

Essential Question How can you show and count 10 objects?

Listen and Draw REAL WORLD

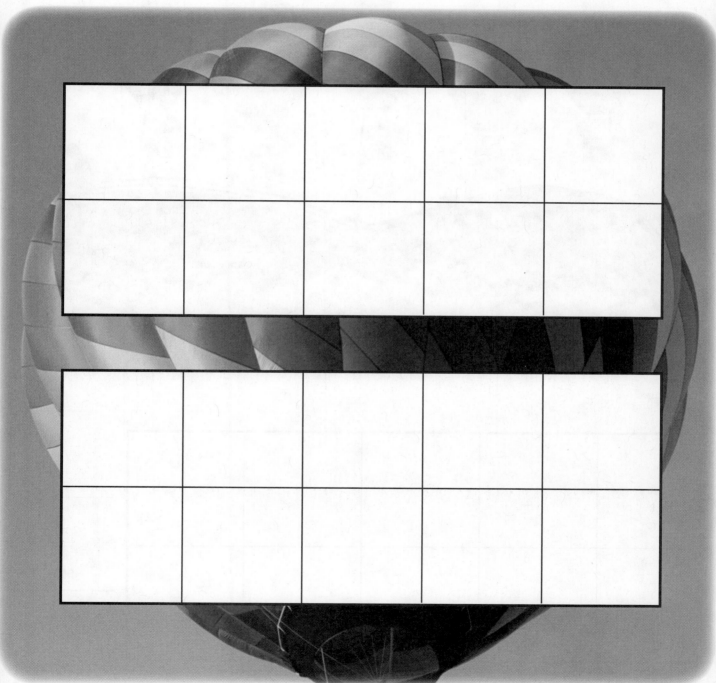

DIRECTIONS Use counters to model 9 in the top ten frame. Use counters to model 10 in the bottom ten frame. Draw the counters. Tell about the ten frames.

Chapter 4 · Lesson 1

one hundred thirty-three **133**

Share and Show

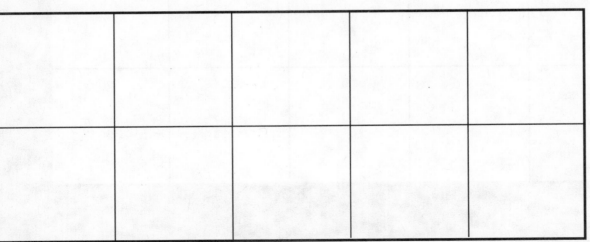

ten

DIRECTIONS 1. Place a counter on each balloon. **2.** Move the counters to the ten frame. Draw the counters. Point to each counter as you count it. Trace the number.

3

10

ten

_____ and _____

_____ and _____

_____ and _____

_____ and _____

© Houghton Mifflin Harcourt Publishing Company

DIRECTIONS **3.** Trace the number. Use counters to model the different ways to make 10. Write to show some pairs of numbers that make 10.

Chapter 4 · Lesson 1

PROBLEM SOLVING

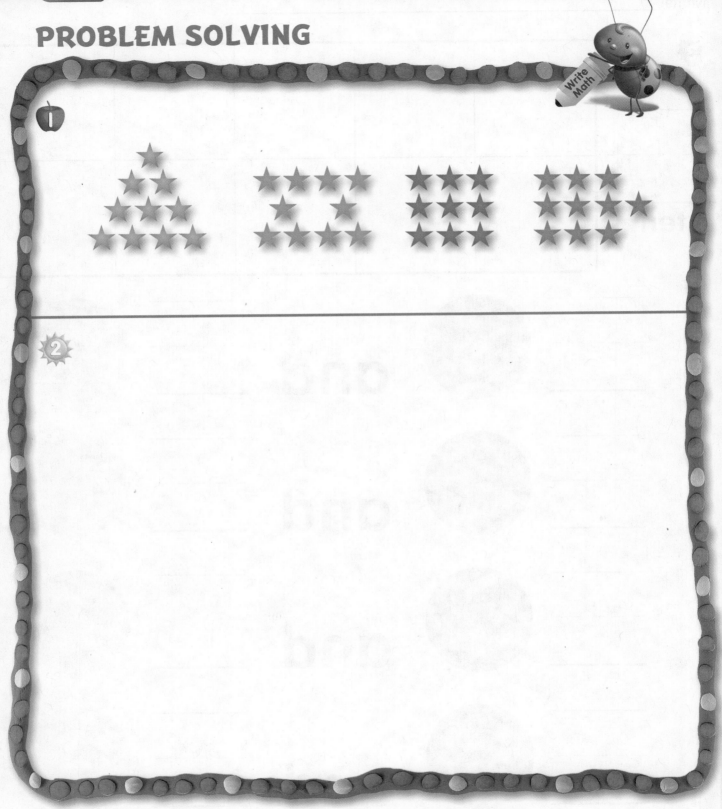

①

②

© Houghton Mifflin Harcourt Publishing Company

DIRECTIONS   1. Count how many in each set. Which sets show ten objects? Circle those sets.   **2.** Draw to show what you know about the number 10. Tell a friend about your drawing.

HOME ACTIVITY • Ask your child to show a set of nine objects. Then have him or her show one more object and tell how many objects are in the set.

FOR MORE PRACTICE: Standards Practice Book, pp. P63–P64

Name _____

Read and Write to 10

Essential Question How can you read and write 10 with words and numbers?

Listen and Draw REAL WORLD

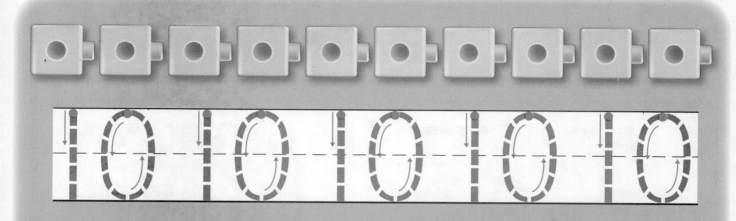

DIRECTIONS Count and tell how many cubes. Trace the numbers. Count and tell how many eggs. Trace the numbers and the word.

Chapter 4 • Lesson 2

Share and Show

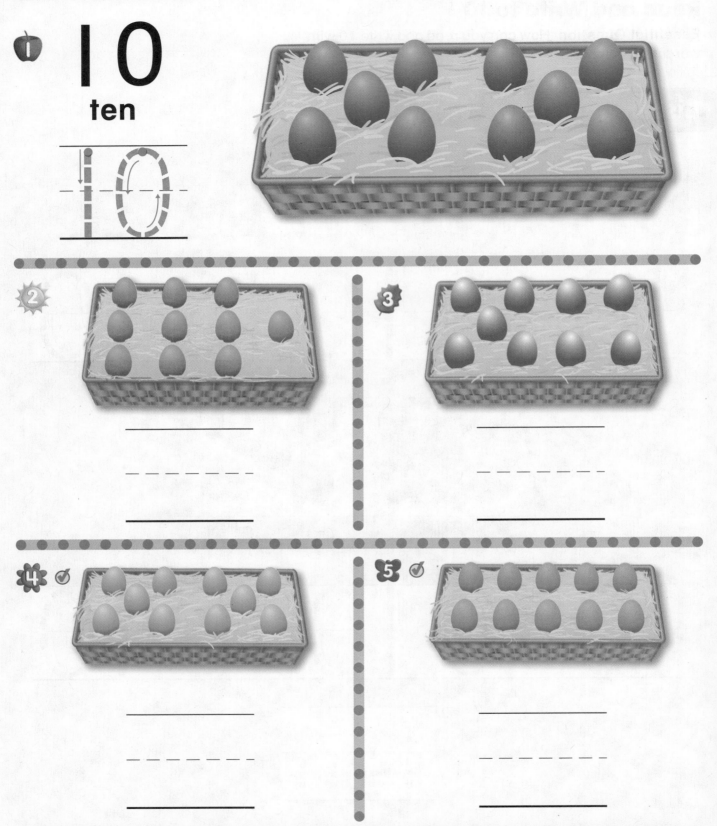

1 **10**
ten

2

3

4 ✓

5 ✓

DIRECTIONS **1.** Count and tell how many eggs. Trace the number. **2–5.** Count and tell how many eggs. Write the number.

6

10
ten

10 10 10 10 10

7

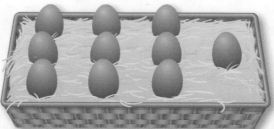

_____

- - - - - - - -

_____

8

_____

- - - - - - - -

_____

9

_____

- - - - - - - -

_____

DIRECTIONS **6.** Say the number. Trace the numbers.
7–9. Count and tell how many. Write the number.

Chapter 4 · Lesson 2

one hundred thirty-nine **139**

PROBLEM SOLVING

DIRECTIONS Draw to show a set that has a number of objects one greater than 9. Write how many objects. Tell a friend about your drawing.

HOME ACTIVITY • Show ten objects. Have your child point to each object in the set as he or she counts them. Then have him or her write the number on paper to show how many objects.

FOR MORE PRACTICE:
Standards Practice Book, pp. P65–P66

Name _____

Algebra • Ways to Make 10

Essential Question How can you use a drawing to make 10 from a given number?

Listen and Draw

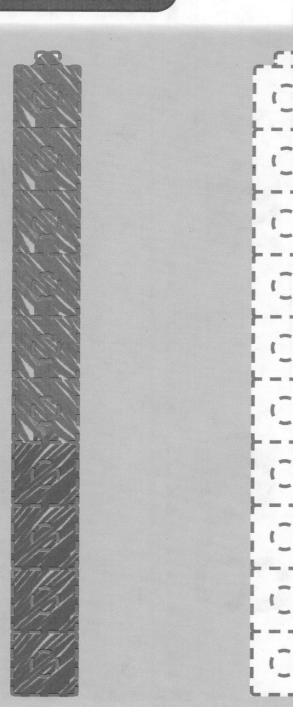

DIRECTIONS Use cubes of two colors to show different ways to make 10. Color to show the ways.

Chapter 4 • Lesson 3

Share and Show

cubes

10

cubes

cubes

9

8

7

DIRECTIONS 1. Count and tell how many cubes of each color there are. Trace the numbers. 2-3. Use blue to color the cubes to match the number. Use red to color the other cubes. Write how many red cubes. Write how many cubes in all.

142 one hundred forty-two

cubes

cubes

cubes

5

3

2

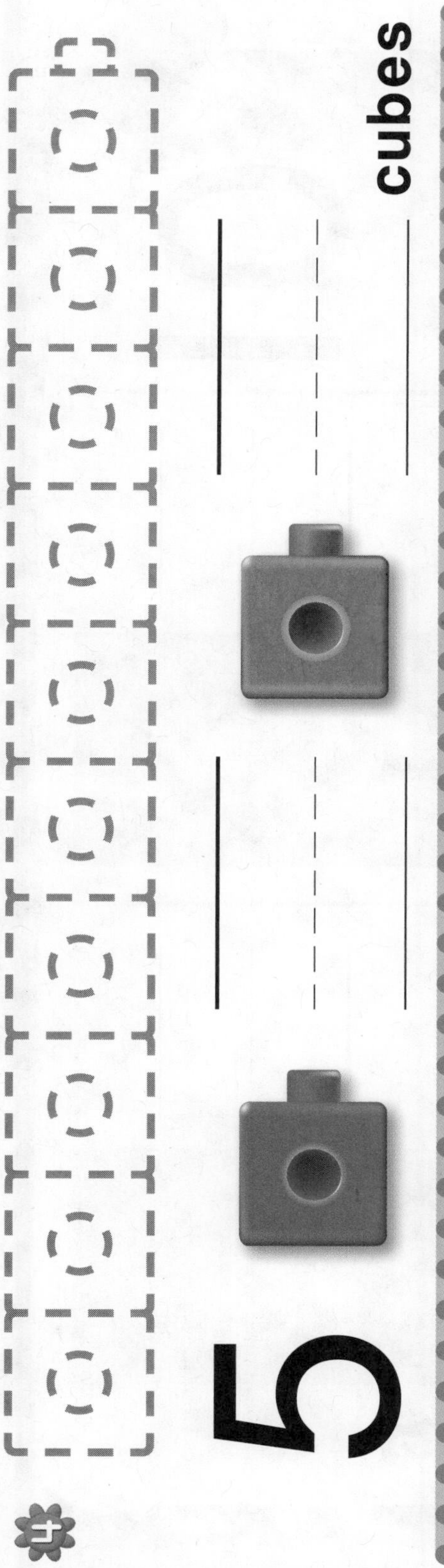

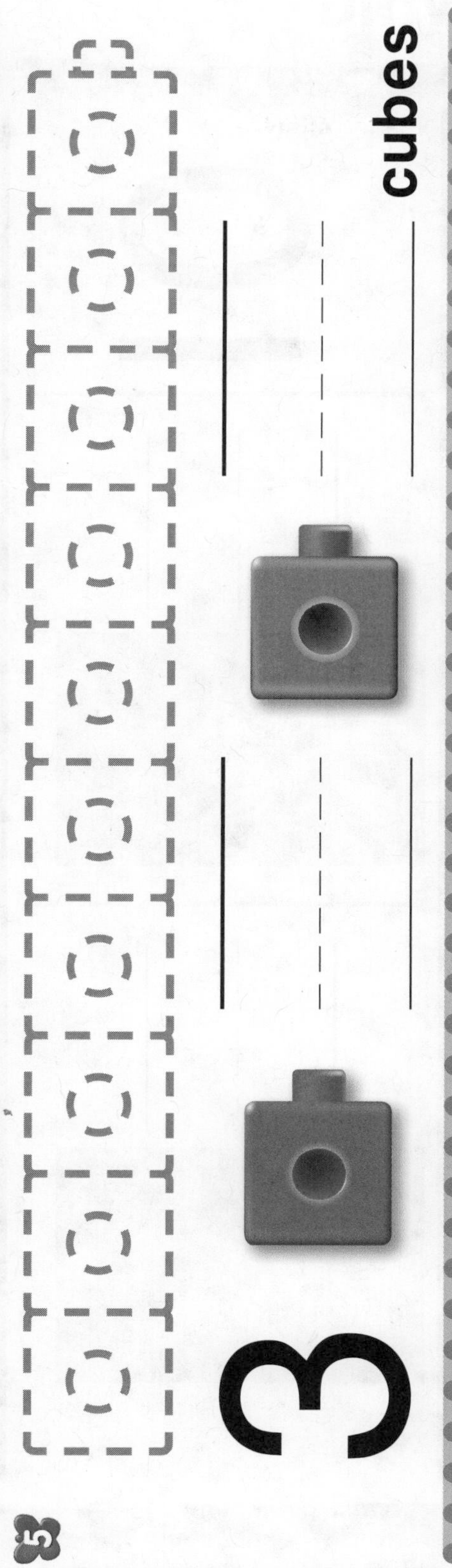

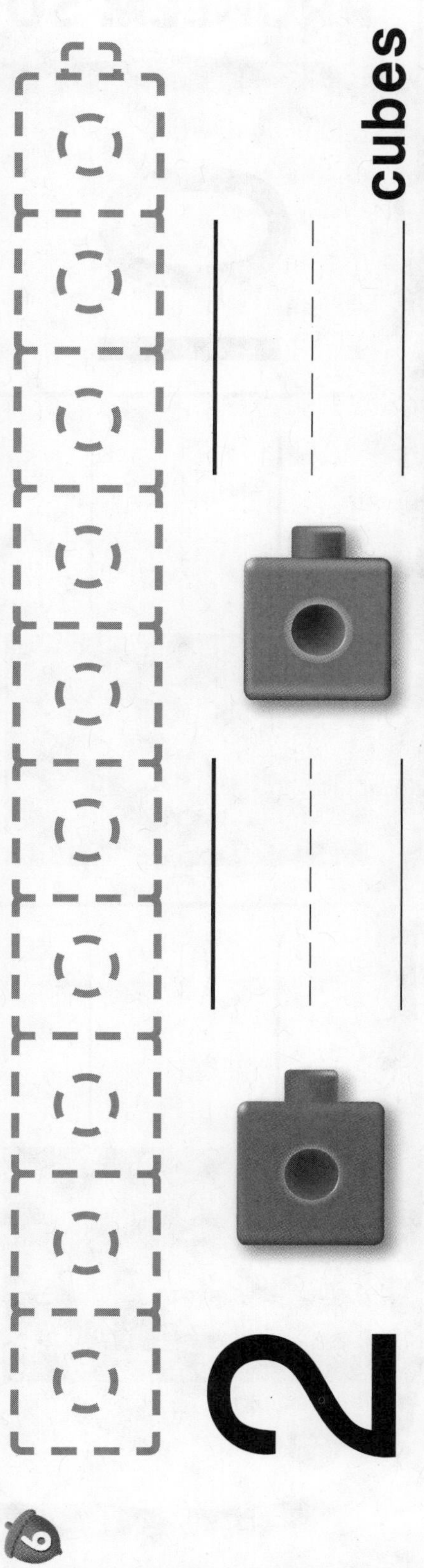

DIRECTIONS **4–6.** Use blue to color the cubes to match the number. Use red to color the other cubes. Write how many red cubes. Write how many cubes in all.

Chapter 4 • Lesson 3

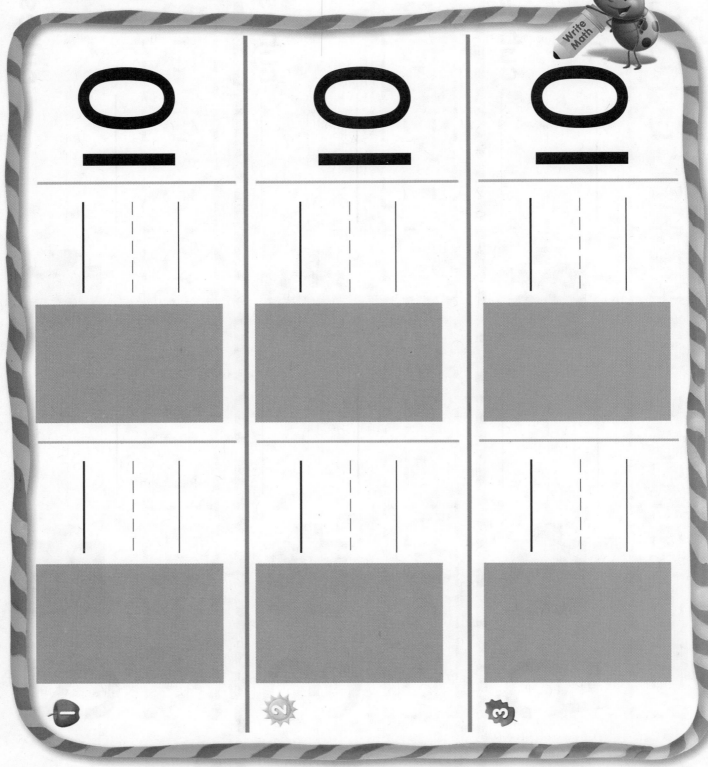

PROBLEM SOLVING

DIRECTIONS 1–3. Use the dot side of two Number Tiles to make 10. Draw the dots. Write the numbers.

HOME ACTIVITY • Ask your child to show a set of 10 objects, using objects of the same kind that are different in one way, for example, large and small paper clips. Then have him or her write the numbers that show how many of each kind are in the set.

FOR MORE PRACTICE:
Standards Practice Book, pp. P67–P68

Name _____

Count and Order to 10

Essential Question How can you count forward to 10 from a given number?

Listen and Draw

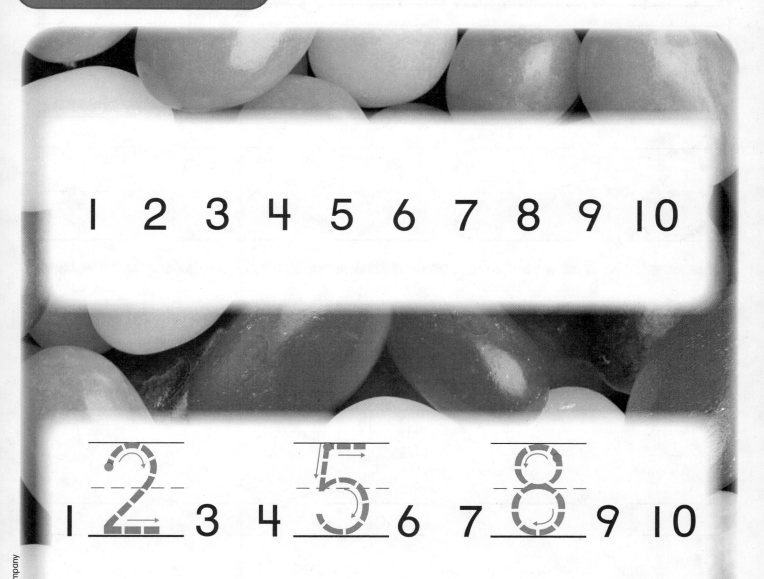

1 2 3 4 5 6 7 8 9 10

1 2 3 4 5 6 7 8 9 10

DIRECTIONS Point to the numbers in the top workspace as you count forward to 10. Trace the numbers in order in the bottom workspace as you count forward to 10.

Share and Show

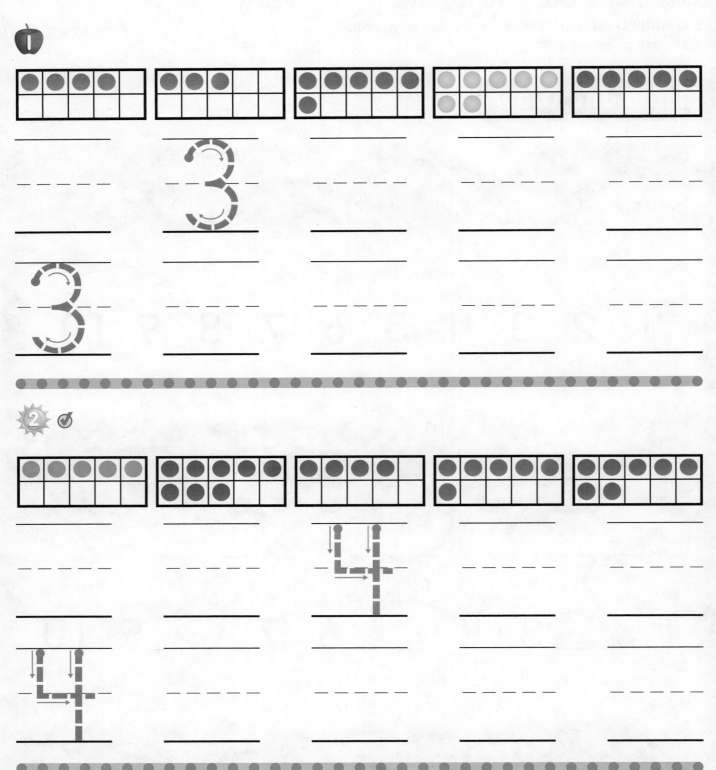

DIRECTIONS 1–2. Count the dots of each color in the ten frames. Trace or write the numbers. Write the numbers in order as you count forward from the dashed number.

3

5

5

4 ✓

6

6

DIRECTIONS 3–4. Count the dots of each color in the ten frames. Trace or write the numbers. Write the numbers in order as you count forward from the dashed number.

HOME ACTIVITY • Write the numbers 1 to 10 in order on a piece of paper. Ask your child to point to each number as he or she counts to 10. Repeat beginning with a number other than 1 when counting.

Chapter 4 • Lesson 4

FOR MORE PRACTICE:
Standards Practice Book, pp. P69–P70

one hundred forty-seven **147**

Concepts and Skills

_____

2

6  cubes

3

     7   8   ___   10

     6   7   8   9
     ○   ○   ○   ○

DIRECTIONS **1.** Place counters in the ten frame to model ten. Draw the counters. Write the number. **2.** Use blue to color the cubes to match the number. Use red to color the other cubes. Write how many red cubes. Write how many cubes in all. **3.** Count forward. Mark under the number that fills the space.

Name _____

Problem Solving • Compare by Matching Sets to 10

Essential Question How can you solve problems using the strategy *make a model*?

🔑 Unlock the Problem

DIRECTIONS Break a ten-cube train into two parts. How can you use matching to compare the parts? Tell a friend about the cube trains. Draw the cube trains.

Try Another Problem

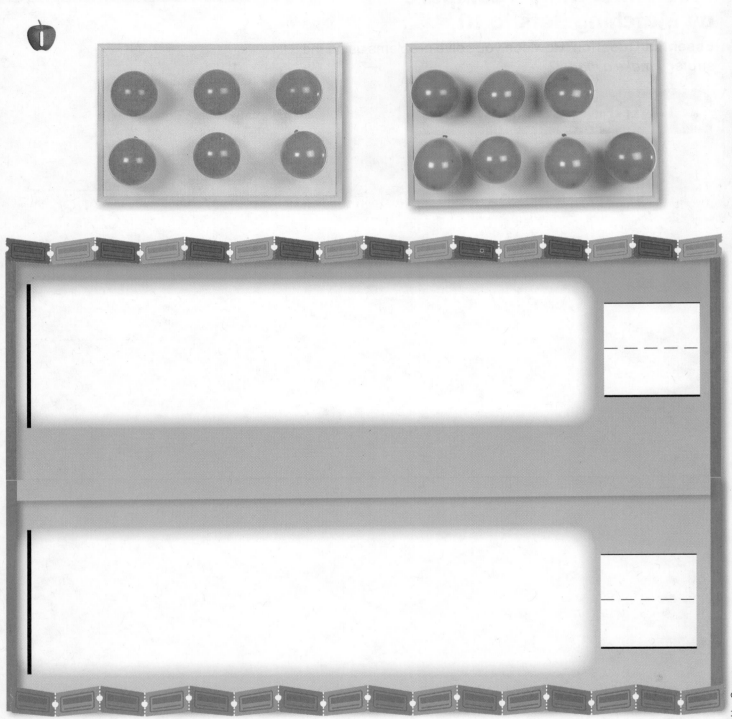

DIRECTIONS 1. Malia has the red balloons. Andrew has the blue balloons. Who has more balloons? Use red and blue cube trains to model the sets of balloons. Compare the cube trains by matching. Draw and color the cube trains. Write how many in each set. Which number is greater? Circle that number.

150 one hundred fifty

Share and Show

② ✓

③

DIRECTIONS **2.** Kyle has 9 tickets. Jared has 7 tickets. Who has fewer tickets? Use cube trains to model the sets of tickets. Compare the cube trains by matching. Draw and color the cube trains. Write how many. Circle the number that is less. **3.** Phil won 8 prizes. Naomi won 5 prizes. Who won fewer prizes? Use cube trains to model the sets of prizes. Compare the cube trains by matching. Draw and color the cube trains. Write how many. Circle the number that is less.

On Your Own

_____

- - - - - - - - -

_____

_____

- - - - - - - - -

_____

DIRECTIONS Look at the model. Are there more blue cubes or more red cubes? Make cube trains of each color. Compare the cube trains by matching. Draw and color the cube trains. Write how many cubes are in each train. Circle the greater number.

HOME ACTIVITY • Ask your child to show two sets of up to 10 objects each. Then have him or her compare the sets by matching and tell which set has more objects.

FOR MORE PRACTICE:
Standards Practice Book, pp. P71–P72

Name _____

Compare by Counting Sets to 10

Essential Question How can you use counting strategies to compare sets of objects?

Listen and Draw REAL WORLD

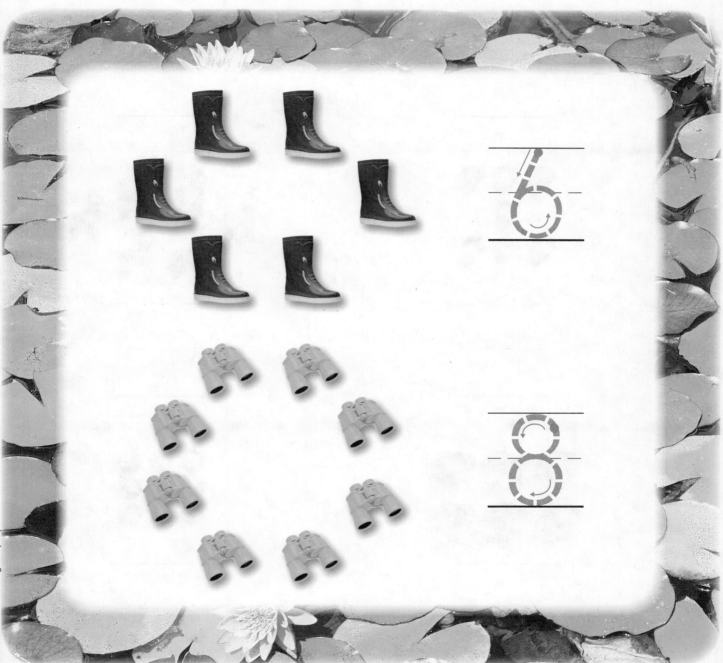

DIRECTIONS Look at the sets of objects. Count how many in each set. Trace the numbers that show how many. Compare the numbers.

Share and Show

 1

_____

- - - - - - - - - -

_____

_____

- - - - - - - - - -

_____

 2 ✓

_____

- - - - - - - - - -

_____

_____

- - - - - - - - - -

_____

 3 ✓

_____

- - - - - - - - - -

_____

_____

- - - - - - - - - -

_____

DIRECTIONS **1–3.** Count how many in each set. Write the number of objects in each set. Compare the numbers. Circle the greater number.

4

_____

– – – – – – – – – – – – – –

_____

_____

– – – – – – – – – – – – – –

_____

5

_____

– – – – – – – – – – – – – –

_____

_____

– – – – – – – – – – – – – –

_____

6

_____

– – – – – – – – – – – – – –

_____

_____

– – – – – – – – – – – – – –

_____

DIRECTIONS **4–6.** Count how many in each set. Write the number of objects in each set. Compare the numbers. Circle the number that is less.

PROBLEM SOLVING REAL WORLD

_____

- - - - - - -

_____

_____

- - - - - - -

_____

DIRECTIONS **1.** Count how many in each set. Write the number of objects in each set. Compare the numbers. Tell a friend about the sets. **2.** Draw to show what you know about counting sets to 10 with the same number of objects.

HOME ACTIVITY • Show your child two sets of up to 10 objects. Have him or her count the objects in each set. Then have him or her compare the numbers of objects in each set, and tell what he or she knows about those numbers.

FOR MORE PRACTICE:
Standards Practice Book, pp. P73–P74

Name _____

Compare Two Numbers

Essential Question How can you compare two numbers between 1 and 10?

Listen and Draw REAL WORLD

7   **7 is less than 8**

           7 is greater than 8

8   **8 is less than 7**

           8 is greater than 7

DIRECTIONS Look at the numbers. As you count forward does 7 come before or after 8? Is it greater or less than 8? Circle the words that describe the numbers when comparing them.

Chapter 4 • Lesson 7   one hundred fifty-seven **157**

Share and Show

1

     3   (8)

2

     10   5

3

     6   4

4 ✓

     7   9

5 ✓

     10   8

DIRECTIONS **I.** Look at the numbers. Think about the counting order as you compare the numbers. Trace the circle around the greater number. **2–5.** Look at the numbers. Think about the counting order as you compare the numbers. Circle the greater number.

158 one hundred fifty-eight

6   2   4

7   5   3

8   8   9

9   10   7

10   6   8

DIRECTIONS **6–10.** Look at the numbers.
Think about the counting order as you
compare the numbers. Circle the number
that is less.

PROBLEM SOLVING REAL WORLD

① _____   _____

‒ ‒ ‒ ‒ ‒   ‒ ‒ ‒ ‒ ‒

_____   _____

② _____   _____

‒ ‒ ‒ ‒ ‒   ‒ ‒ ‒ ‒ ‒

_____   _____

DIRECTIONS **I.** John has a number of apples that is greater than 5 and less than 7. Cody has a number of apples that is two less than 8. Write how many apples each boy has. Compare the numbers. Tell a friend about the numbers. **2.** Write two numbers between I and I0. Tell a friend about the two numbers.

HOME ACTIVITY • Write the numbers I to I0 on individual pieces of paper. Select two numbers and ask your child to compare the numbers and tell which number is greater and which number is less.

160 one hundred sixty

FOR MORE PRACTICE:
Standards Practice Book, pp. P75–P76

Chapter 4 Review/Test

Vocabulary

①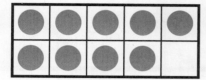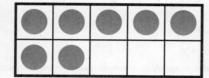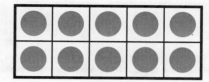

ten

Concepts and Skills

②

_____
- - - - - -
_____

③

cubes

© Houghton Mifflin Harcourt Publishing Company

DIRECTIONS 1. Draw a line to match the number word to the counters in the ten frame. (pp. 133–136) 2. Use counters to model the number 10. Draw the counters. Write the number. 3. Use blue to color the cubes to match the number. Use red to color the other cubes. Write how many red cubes. Write how many cubes in all.

4

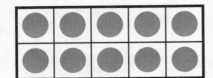

7   8   9   10

○   ○   ○   ○

5

10

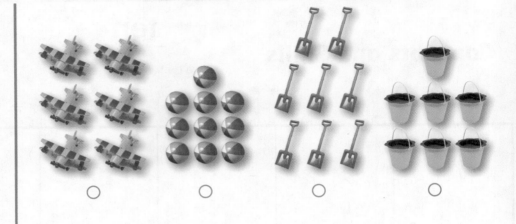

○   ○   ○   ○

6

7

5   6   7   _   9   10

6   8   9   10

○   ○   ○   ○

DIRECTIONS **4.** Mark under the number that shows how many counters at the beginning of the row. **5.** Mark under the set that models the number at the beginning of the row. **6.** Mark beside the cube train that shows a way to make 10. **7.** Count forward. Mark under the number that fills the space.

8

○

○

○

○

9

○

○

○

○

10

7 | 5   6   7   8

       ○   ○   ○   ○

11

7 | 6   7   8   9

       ○   ○   ○   ○

DIRECTIONS **8.** Compare the cube trains by matching. Mark beside the cube train that has a greater number of cubes. **9.** Compare the sets by counting. Mark beside the set that has a number of objects that is less than the number of objects in the other sets. **10.** Mark under the number that is greater than the number at the beginning of the row. **11.** Mark under the number that is less than the number at the beginning of the row.

Performance Task

© Houghton Mifflin Harcourt Publishing Company

DIRECTIONS This task will assess the child's understanding of representing and comparing numbers to 10.

Curious About Math with Curious George

Most ladybugs have red, orange, or yellow wing covers and black spots.

• How many ladybugs do you see?

Name _____

Show What You Know ✓

More

 1

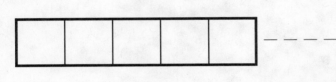

 2

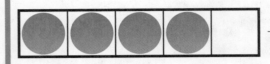

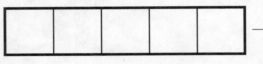

Compare Numbers to 10

 3

DIRECTIONS **1–2.** Count and tell how many. Draw a set with one more counter. Write how many in each set. **3.** Write the number of cubes in each set. Circle the number that is greater than the other number.

FAMILY NOTE: This page checks your child's understanding of important skills needed for success in Chapter 5.

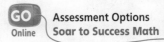
GO Online · Assessment Options · Soar to Success Math

Name _____

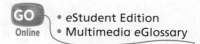

ten

DIRECTIONS Count and tell how many birds are on the ground. Count and tell how many birds are flying. Write these numbers to show a pair of numbers that make ten.

GO Online · eStudent Edition · Multimedia eGlossary

Game

Pairs That Make 7

DIRECTIONS Play with a partner. The first player rolls the number cube and writes the number on the yellow boat. Partners determine what number makes 7 when paired with the number on the yellow boat. Players take turns rolling the number cube until that number is rolled. Write the number beside it on the green boat. Partners continue to roll the number cube finding pairs of numbers that make 7.

MATERIALS number cube (1–6)

Addition: Add To

Essential Question How can you show addition as adding to?

 **Listen and Draw** REAL WORLD

DIRECTIONS Listen to the addition word problem. Trace the number that shows how many children are on the swings. Trace the number that shows how many children are being added to the group. Trace the number that shows how many children there are now.

 **and** _____

_____

DIRECTIONS 1. Listen to the addition word problem. Trace the number that shows how many children are sitting eating lunch. Write the number that shows how many children are being added to the group. Write the number that shows how many children are having lunch now.

170 one hundred seventy

Name _____

_____   _____

and _____

_____

DIRECTIONS   2. Listen to the addition word problem. Write the
number that shows how many children are playing with the soccer ball.
Write the number that shows how many children are being added to the
group. Write the number that shows how many children there are now.

Chapter 5 • Lesson 1   one hundred seventy-one **171**

PROBLEM SOLVING REAL WORLD

1

——— **and** ———

2

———

DIRECTIONS   1. Two sheep are in a pen. Two sheep are added to the pen. How can you write the numbers to show the sheep being added? **2.** Write how many sheep are in the pen now.

HOME ACTIVITY · Show your child a set of four objects. Have him or her add one object to the set and tell how many there are now.

FOR MORE PRACTICE:
Standards Practice Book, pp. P81–P82

Addition: Put Together

Essential Question How can you show addition as putting together?

Listen and Draw

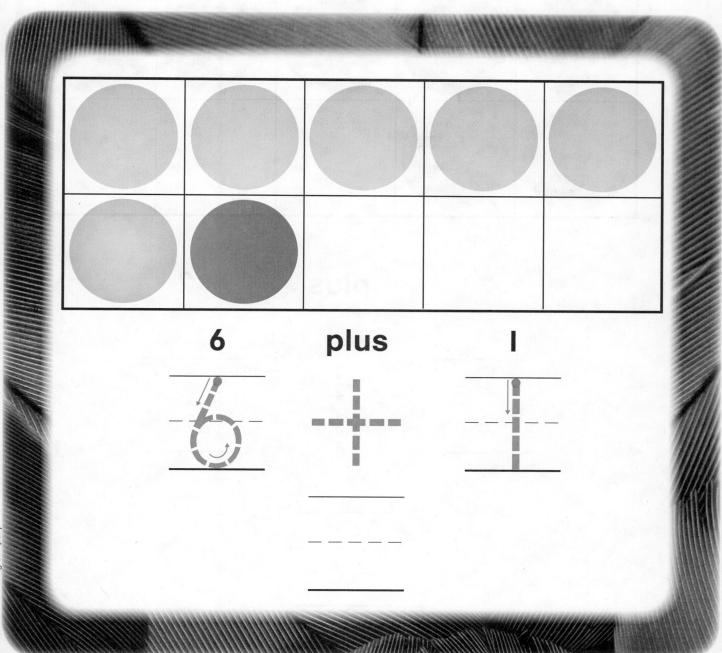

6   **plus**   **l**

DIRECTIONS Listen to the addition word problem. Place red and yellow counters in the ten frame as shown. Trace the numbers and the symbol to show the sets that are put together. Write the number that shows how many in all.

Share and Show

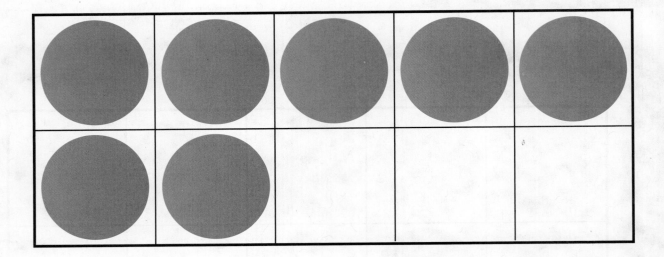

7   **plus**   **2**

_____   _____
- - - - -   ✚   - - - - -
_____   _____

_____
- - - - -
_____

DIRECTIONS 1. Listen to the addition word problem. Place red counters in the ten frame as shown. Place yellow counters to model the sets that are put together. Write the numbers and trace the symbol. Write the number to show how many in all.

174 one hundred seventy-four

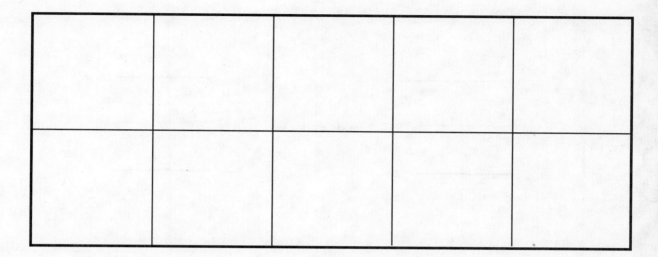

2   **plus**   **8**

_____   ┼   _____

- - - - - - -   - - - - - - -

_____   _____

_____

- - - - - - -

_____

DIRECTIONS **2.** Listen to the addition word problem. Place counters in the ten frame to model the sets that are put together. How many are there of each color counter? Write the numbers and trace the symbol. Write the number to show how many in all.

PROBLEM SOLVING REAL WORLD

DIRECTIONS **1.** Four red apples and two green apples are on the table. Write the numbers and trace the symbol to show the apples being put together. **2.** Write the number to show how many apples in all.

HOME ACTIVITY • Show your child two sets of 4 objects. Have him or her put the sets of objects together and tell how many in all.

FOR MORE PRACTICE: Standards Practice Book, pp. P83–P84

Name _____

Problem Solving • Act Out Addition Problems

Essential Question How can you solve problems using the strategy *act it out*?

Unlock the Problem REAL WORLD

DIRECTIONS Listen to and act out the addition word problem. Trace the addition sentence. Tell a friend how many children in all.

Chapter 5 • Lesson 3

one hundred seventy-seven **177**

Try Another Problem

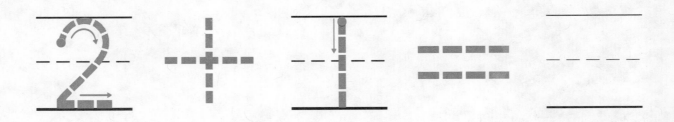

DIRECTIONS **1.** Listen to and act out the addition word problem. Trace the numbers and the symbols. Write the number that shows how many children in all.

178 one hundred seventy-eight

Share and Show

 $+$  $=$

• •

DIRECTIONS **2.** Listen to and act out the addition word problem. Trace the numbers and the symbols. Write the number that shows how many children in all.

On Your Own  REAL WORLD

 3 + 1 = ___

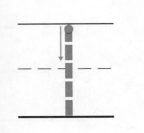 1 + 4 = ___

DIRECTIONS 1. Tell an addition word problem about the puppies. Trace the numbers and the symbols. Write the number that shows how many puppies there are now. 2. Draw a picture to match this addition sentence. Write how many in all. Tell a friend about your drawing.

HOME ACTIVITY • Tell your child a short word problem about adding three objects to a set of two objects. Have your child use toys to act out the word problem.

180 one hundred eighty

FOR MORE PRACTICE:
Standards Practice Book, pp. P85–P86

Name _____

Algebra • Model and Draw
Addition Problems

Essential Question How can you use objects and drawings to solve addition word problems?

Listen and Draw

DIRECTIONS Place cubes as shown. Listen to the addition word problem. Model to show the cubes put together in a cube train. Color to show how the cube train looks. Trace to complete the addition sentence.

Chapter 5 • Lesson 4

Share and Show

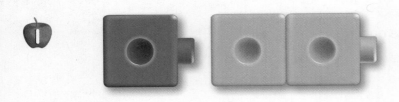

DIRECTIONS 1–2. Place cubes as shown. Listen to the addition word problem. Model to show the cubes put together. Draw the cube train. Trace and write to complete the addition sentence.

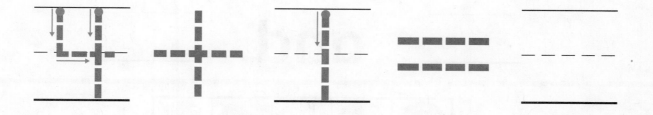

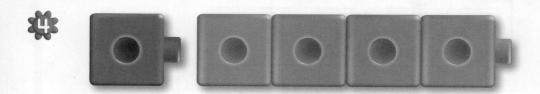

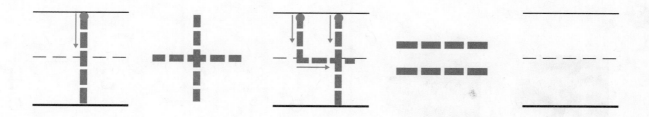

DIRECTIONS **3–4.** Place cubes as shown. Listen to the addition word problem. Model to show the cubes put together. Draw the cube train. Trace and write to complete the addition sentence.

HOME ACTIVITY • Tell your child a short word problem about adding two objects to a set of another two objects. Have him or her use objects to model the word problem.

Chapter 5 • Lesson 4

FOR MORE PRACTICE:
Standards Practice Book, pp. P87–P88

one hundred eighty-three **183**

Concepts and Skills

_____   **and**   _____

- - - - - - -   - - - - -

_____   _____

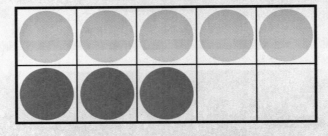

_____   **+**   _____

- - - - - - -   - - - - -

_____   _____

1	2	3	4
○	○	○	○

DIRECTIONS **1.** Write the number that shows how many puppies are sitting. Write the number that shows how many puppies are being added to them. **2.** Write the numbers and trace the symbol to show the sets that are put together. **3.** Mark under the number that shows how many cubes there are when these are put together.

Algebra • Write Addition Sentences for 10

Essential Question How can you use a drawing to find the number that makes a ten from a given number?

Listen and Draw

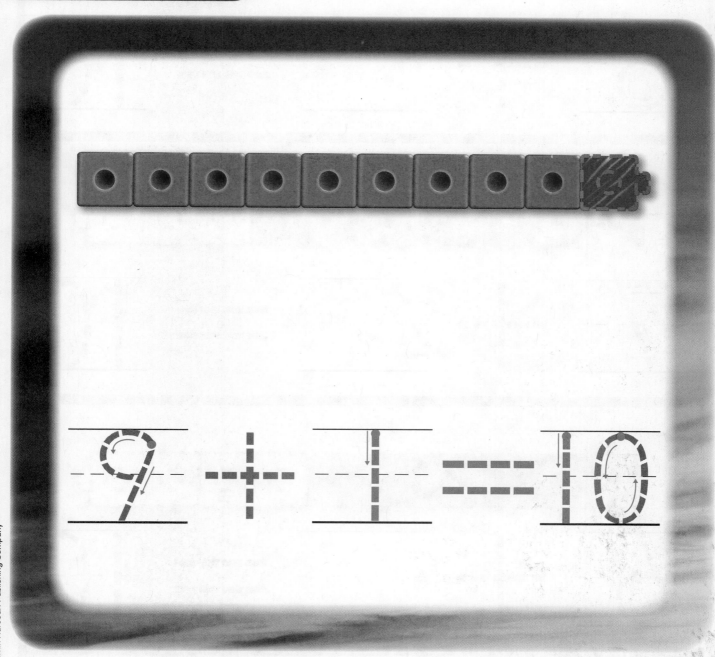

DIRECTIONS Look at the cube train. How many red cubes do you see? How many blue cubes do you need to add to make 10? Trace the blue cube. Trace to show this as an addition sentence.

Share and Show

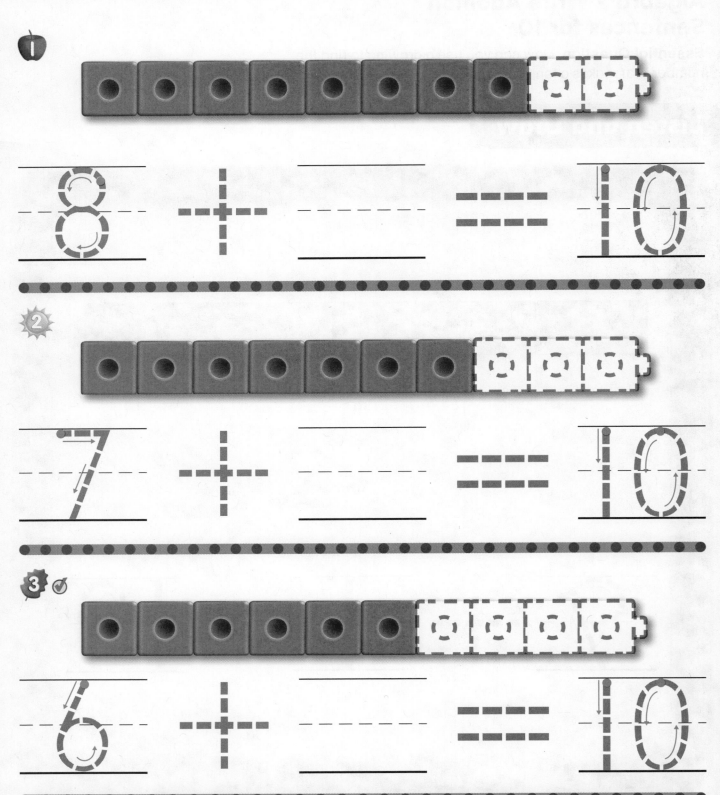

1 8 + ___ = 10

2 7 + ___ = 10

3 ✓ 6 + ___ = 10

DIRECTIONS 1-3. Look at the cube train. How many red cubes do you see? How many blue cubes do you need to add to make 10? Use blue to color those cubes. Write and trace to show this as an addition sentence.

❀ 4

5 + ___ = 10

✤ 5

4 + ___ = 10

🌰 6

3 + ___ = 10

DIRECTIONS 4–6. Look at the cube train. How many red cubes do you see? How many blue cubes do you need to add to make 10? Use blue to draw those cubes. Write and trace to show this as an addition sentence.

PROBLEM SOLVING REAL WORLD

1

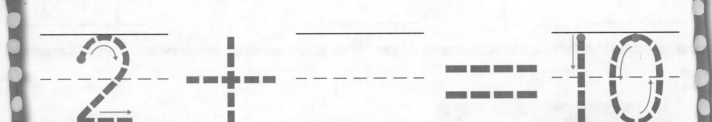

$$2 + \underline{\hspace{3cm}} = 10$$

2

$$1 + \underline{\hspace{3cm}} = 10$$

DIRECTIONS **1.** Troy has 2 ducks. How many more ducks does he need to get to have 10 ducks in all? Draw to solve the problem. Trace and write to show this as an addition sentence. **2.** Draw to find the number that makes 10 when put together with the given number. Trace and write to show this as an addition sentence.

HOME ACTIVITY • Show your child a number from 1 to 9. Ask him or her to find the number that makes 10 when put together with that number. Then have him or her tell a story to go with the problem.

188 one hundred eighty-eight

FOR MORE PRACTICE:
Standards Practice Book, pp. P89–P90

Name _____

Algebra • Write Addition Sentences

Essential Question How can you solve addition word problems and complete the addition sentence?

Listen and Draw REAL WORLD

$$2 + 1 = 3$$

DIRECTIONS Listen to the addition word problem. How many are in the set you start with. How many are being added to the set? How many are there now? Trace the addition sentence.

Share and Show

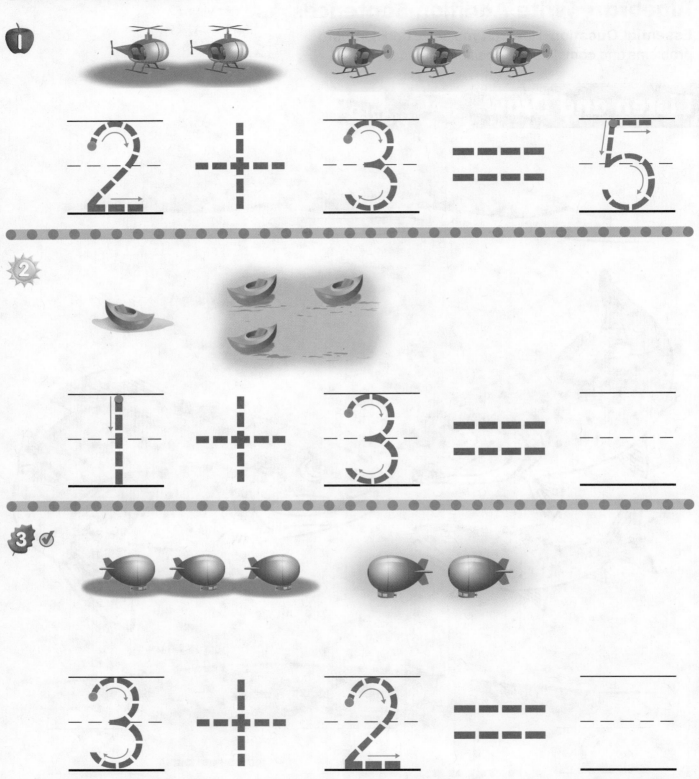

1. $2 + 3 = 5$

2. $1 + 3 = \underline{}$

3. $3 + 2 = \underline{}$

DIRECTIONS **1.** Listen to the addition word problem. How many are in the set you start with. How many are being added to the set? How many are there now? Trace the addition sentence. **2–3.** Listen to the addition word problem. How many are in the set you start with. How many are being added to the set? How many are there now? Trace and write the numbers to complete the addition sentence.

Name _____

4

1 + 4 = ___

5

3 + 1 = ___

6

2 + 3 = ___

© Houghton Mifflin Harcourt Publishing Company

DIRECTIONS   4–6. Tell an addition word problem about the sets. How many are in the set you start with. How many are being added to the set? How many are there now? Trace and write the numbers to complete the addition sentence.

PROBLEM SOLVING REAL WORLD

1

2

DIRECTIONS **I.** Bill catches two fish. Jake catches two fish. How many fish do they catch in all? Draw to show the fish. Trace and write to complete the addition sentence. **2.** Tell a different addition word problem about fish. Draw to show the fish. Tell about your drawing. Complete the addition sentence.

HOME ACTIVITY • Have your child show three fingers on one hand. Have him or her show two fingers on the other hand. Then have him or her tell how many fingers he or she showed in all.

FOR MORE PRACTICE: Standards Practice Book, pp. P91–P92

Algebra • Write More Addition Sentences

Essential Question How can you solve addition word problems and complete the addition sentence?

Listen and Draw REAL WORLD

$$4 + 1 = 5$$

DIRECTIONS Listen to the addition word problem about the birds. How many birds did you start with? How many birds are joining? How many birds are there now? Trace the addition sentence.

Share and Show

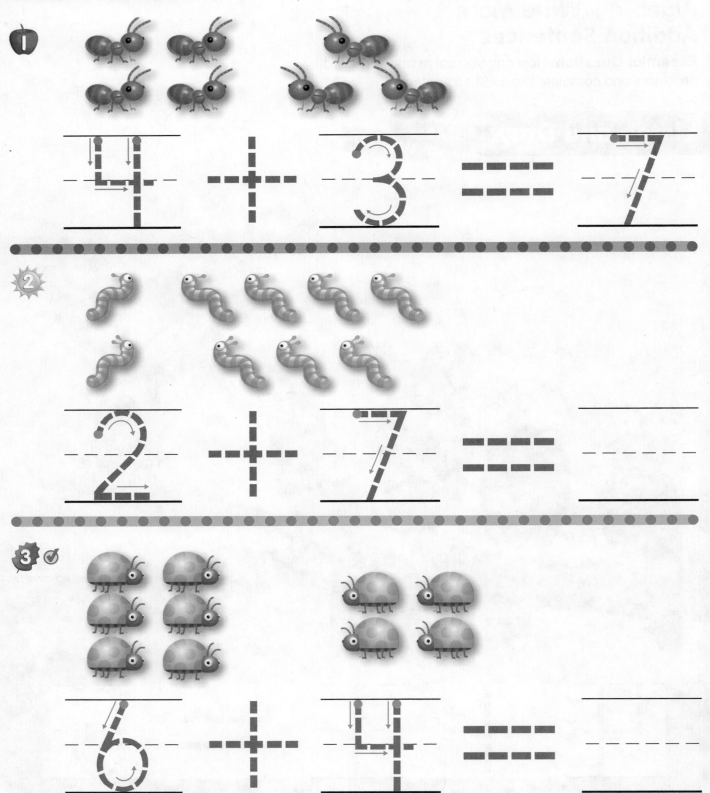

1 4 + 3 == 7

2 2 + 7 ==

3 ✓ 6 + 4 ==

DIRECTIONS **1.** Listen to the addition word problem. How many are in the set to start with? How many are joining? How many are there now? Trace to complete the addition sentence. **2–3.** Listen to the addition word problem. How many are in the set to start with? How many are joining? How many are there now? Trace and write to complete the addition sentence.

🌼 **4**

$$3 + 5 = \underline{\hspace{3cm}}$$

🦋 **5**

$$6 + 3 = \underline{\hspace{3cm}}$$

🌰 **6**

$$2 + 8 = \underline{\hspace{3cm}}$$

DIRECTIONS   4–6. Tell an addition word problem. How many are in the set to start with? How many are joining? How many are there now? Trace and write to complete the addition sentence.

PROBLEM SOLVING

$$7 = 5 + 2$$

DIRECTIONS **1.** Read the addition sentence. Draw to show what you know about the addition sentence. Tell a friend about your drawing.

HOME ACTIVITY · Tell your child an addition word problem such as: There are six socks in the drawer. I added four more socks. How many socks are in the drawer now?

FOR MORE PRACTICE:
Standards Practice Book, pp. P93–P94

Name _____

Algebra • Number Pairs to 5

Essential Question How can you solve problems using number pairs for sums to 5?

Listen and Draw

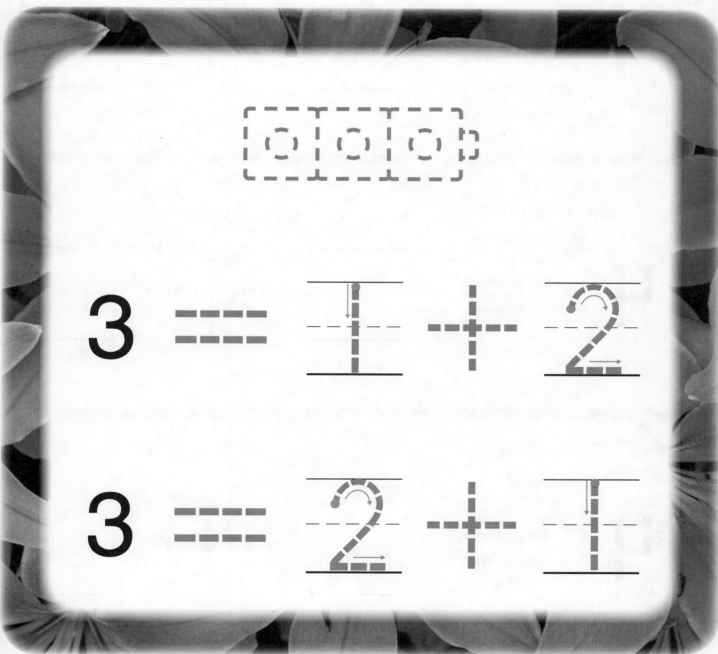

$$3 = 1 + 2$$

$$3 = 2 + 1$$

DIRECTIONS Listen to the word problem. Place two colors of cubes on the cube train to show the number pairs that make 3. Trace the addition sentences to show some of the number pairs.

Chapter 5 • Lesson 8

© Houghton Mifflin Harcourt Publishing Company

Share and Show

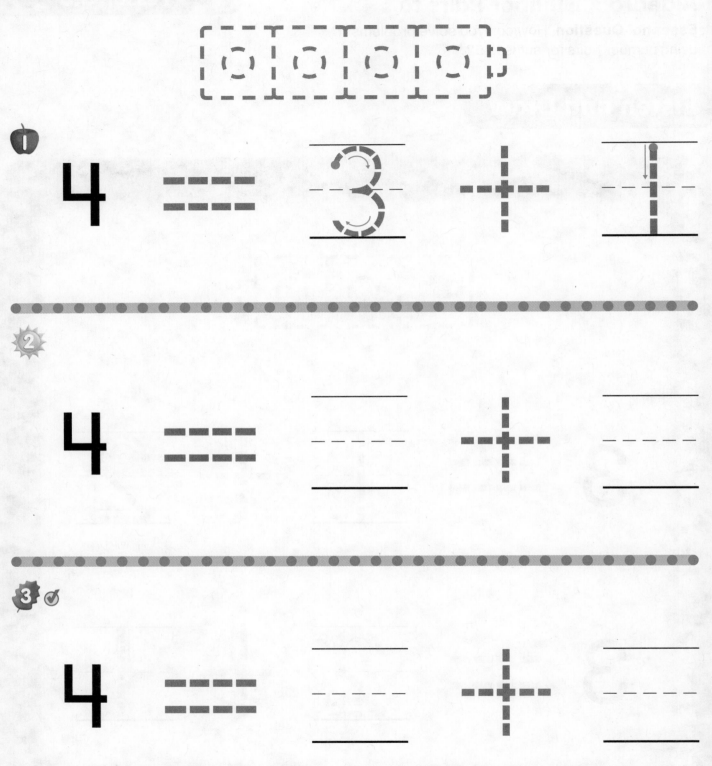

1. 4 = 3 + 1

2. 4 = ___ + ___

3. 4 = ___ + ___

© Houghton Mifflin Harcourt Publishing Company

DIRECTIONS Listen to the word problem. Place two colors of cubes on the cube train to show the number pairs that make 4. **1.** Trace the addition sentence to show one of the pairs. **2–3.** Complete the addition sentence to show another number pair. Color the cube train to match the addition sentence in Exercise 3.

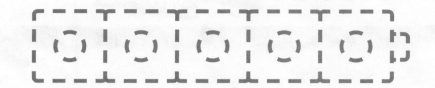

4 5

5 5

6 5

7 5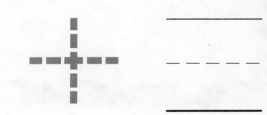

DIRECTIONS Listen to the word problem. Place two colors of cubes on the cube train to show the number pairs that make 5. **4–7.** Complete the addition sentence to show a number pair. Color the cube train to match the addition sentence in Exercise 7.

PROBLEM SOLVING REAL WORLD

$$5 = \underline{} + \underline{}$$

DIRECTIONS **1.** Peyton and Ashley have five red apples. Peyton is holding five of the apples. How many is Ashley holding? Color the cube train to show the number pair. Complete the addition sentence. **2.** Draw to show what you know about a number pair to 5.

HOME ACTIVITY • Have your child tell you the number pairs for a set of objects up to five. Have him or her tell an addition sentence for one of the number pairs.

FOR MORE PRACTICE:
Standards Practice Book, pp. P95–P96

Name _____

Algebra • Number Pairs for 6 and 7

Essential Question How can you solve problems using number pairs for each sum of 6 and 7?

Listen and Draw

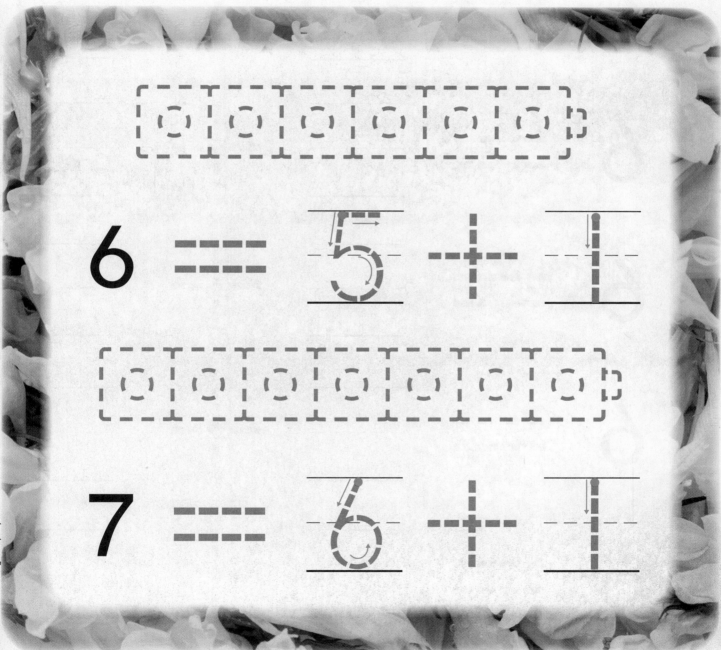

DIRECTIONS Listen to the word problem. Place two colors of cubes on the cube trains to match the addition sentences. Color the cube trains. Trace the addition sentences.

Chapter 5 • Lesson 9

two hundred one **201**

Share and Show

1 6 === $1 + 5$

2 6 === ___ + ___

3 6 === ___ + ___

4 ✓ 6 === ___ + ___

5 ✓ 6 === ___ + ___

DIRECTIONS Listen to the word problem. Place two colors of cubes on the cube train to show the number pairs that make 6. **1.** Trace the addition sentence to show one of the pairs. **2–5.** Complete the addition sentence to show a number pair for 6. Color the cube train to match the addition sentence in Exercise 5.

Name _____

6. 7

7. 7

8. 7

9. 7

10. 7

DIRECTIONS Listen to the word problem. Place two colors of cubes on the cube train to show the number pairs that make 7. **6–10.** Complete the addition sentence to show a number pair for 7. Color the cube train to match the addition sentence in Exercise 10.

Mathematical Practices • **Model • Reason • Make Sense**

PROBLEM SOLVING

 6 === ___ ___ + ___ ___

 7 === ___ ___ + ___ ___

DIRECTIONS 1. Peter and Grant have six toy cars. Peter has no cars. How many cars does Grant have? Color the cube train to show the number pair. Complete the addition sentence. **2.** Draw to show what you know about a number pair for 7 when one number is 0. Complete the addition sentence.

HOME ACTIVITY • Have your child use his or her fingers on two hands to show a number pair for 6.

FOR MORE PRACTICE: Standards Practice Book, pp. P97–P98

Name _____

Algebra • Number Pairs for 8

Essential Question How can you solve problems using number pairs for sums of 8?

Listen and Draw

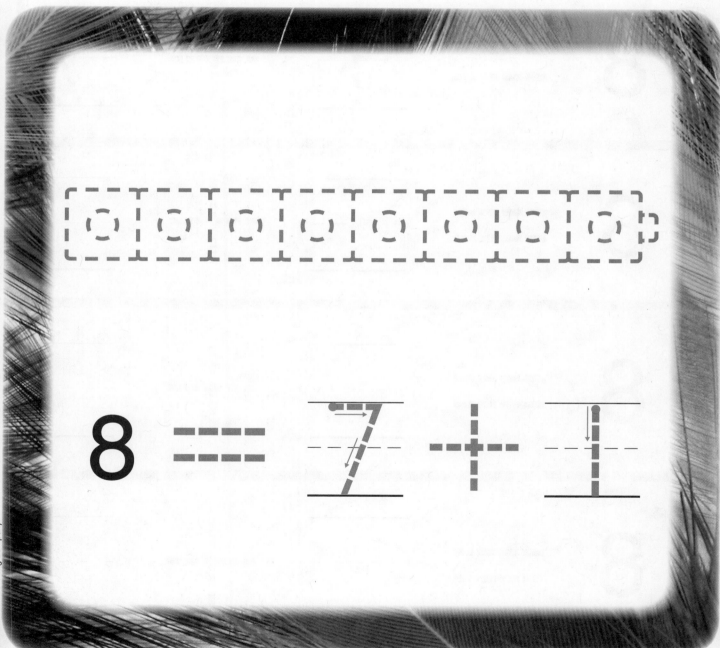

$$8 = 7 + 1$$

DIRECTIONS Listen to the word problem. Use two colors of cubes to make a cube train to match the addition sentence. Color the cube train to show your work. Trace the addition sentence.

Share and Show

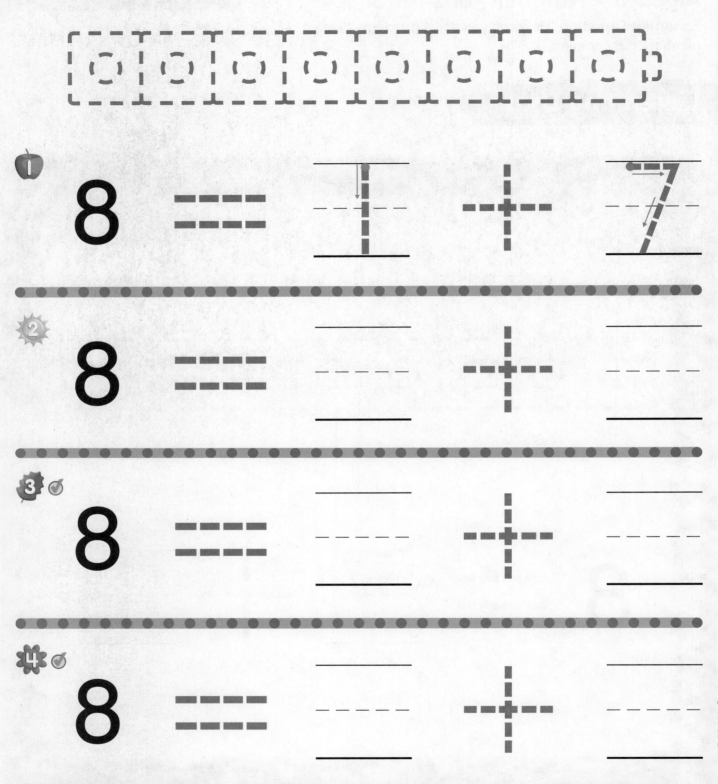

1. 8 === 1 + 7

2. 8 === __ + __

3. 8 === __ + __

4. 8 === __ + __

DIRECTIONS Listen to the word problem. Use two colors of cubes to make a cube train to show the number pairs that make 8. **1.** Trace the addition sentence to show one of the pairs. **2–4.** Complete the addition sentence to show a number pair for 8. Color the cube train to match the addition sentence in Exercise 4.

Name _____

8

___ + ___

8

8 ===

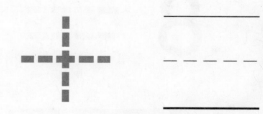

DIRECTIONS Listen to the word problem. Use two colors of cubes to make a cube train to show the number pairs that make 8. 5–7. Complete the addition sentence to show a number pair for 8. Color the cube train to match the addition sentence in Exercise 7.

Chapter 5 · Lesson 10   two hundred seven **207**

PROBLEM SOLVING REAL WORLD

1

8 ==== _____ _____ + _____
          _____   _____

2

8 ==== _____ _____ + _____
          _____   _____

DIRECTIONS 1. There are eight crayons in a packet. Eight of the crayons are red. How many are not red? Draw and color to show how you solved. Complete the addition sentence. **2.** Draw to show what you know about a different number pair for 8. Complete the addition sentence.

HOME ACTIVITY • Have your child tell you the number pairs for a set of eight objects. Have him or her tell the addition sentence to match one of the number pairs.

FOR MORE PRACTICE:
Standards Practice Book, pp. P99–P100

Name _____

Algebra • Number Pairs for 9

Essential Question How can you solve problems using number pairs for sums of 9?

Listen and Draw

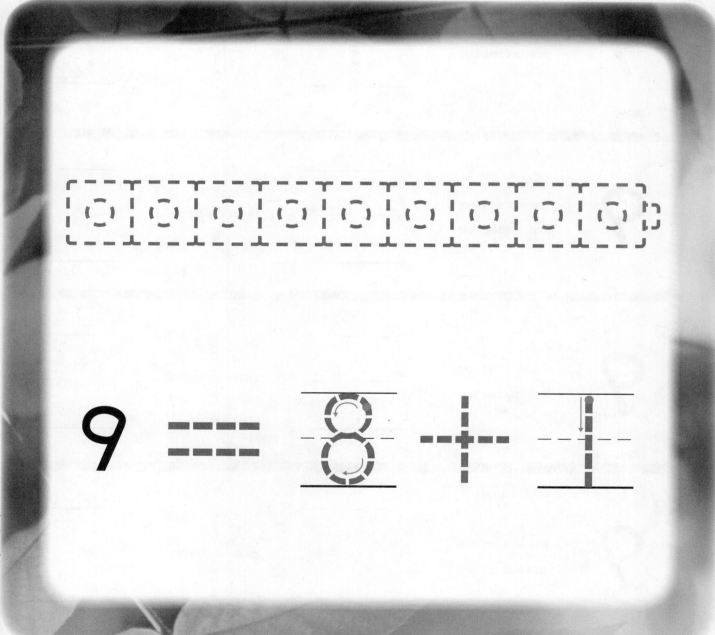

$$9 = 8 + 1$$

DIRECTIONS Listen to the word problem. Use two colors of cubes to make a cube train to match the addition sentence. Color the cube train to show your work. Trace the addition sentence.

Chapter 5 • Lesson 11 two hundred nine **209**

Share and Show

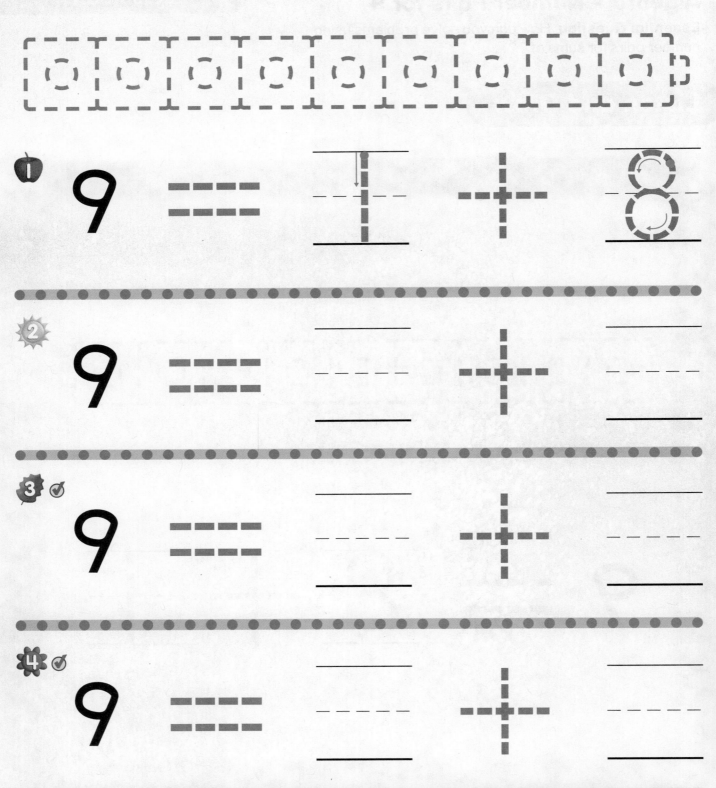

1 9 = _ _ _ / + / 8

2 9 = _ _ _ ____ + ____

3 ✓ 9 = _ _ _ ____ + ____

4 ✓ 9 = _ _ _ ____ + ____

DIRECTIONS Listen to the word problem. Use two colors of cubes to make a cube train to show the number pairs that make 9. **1.** Trace the addition sentence to show one of the pairs. **2–4.** Complete the addition sentence to show a number pair for 9. Color the cube train to match the addition sentence in Exercise 4.

5

9   = = =
= = =

_____   +   _____

6

9   = =
= = =

_____   +   _____

7

9   = = =
= = =

_____   +   _____

8

9   = = =
= = =

_____   +   _____

DIRECTIONS Listen to the word problem. Use two colors of cubes to make a cube train to show the number pairs that make 9. **5–8.** Complete the addition sentence to show a number pair for 9. Color the cube train to match the addition sentence in Exercise 8.

PROBLEM SOLVING REAL WORLD

1

9 = _____ _____ + _____

2

9 = _____ + _____

DIRECTIONS 1. Shelby has nine friends. None of them are boys. How many are girls? Complete the addition sentence to show the number pair. **2.** Draw to show what you know about a different number pair for 9. Complete the addition sentence.

HOME ACTIVITY · Have your child use his or her fingers on two hands to show a number pair for 9.

212 two hundred twelve

FOR MORE PRACTICE:
Standards Practice Book, pp. P101–P102

Name _____

Algebra • Number Pairs for 10

Essential Question How can you solve problems using number pairs for sums of 10?

Listen and Draw

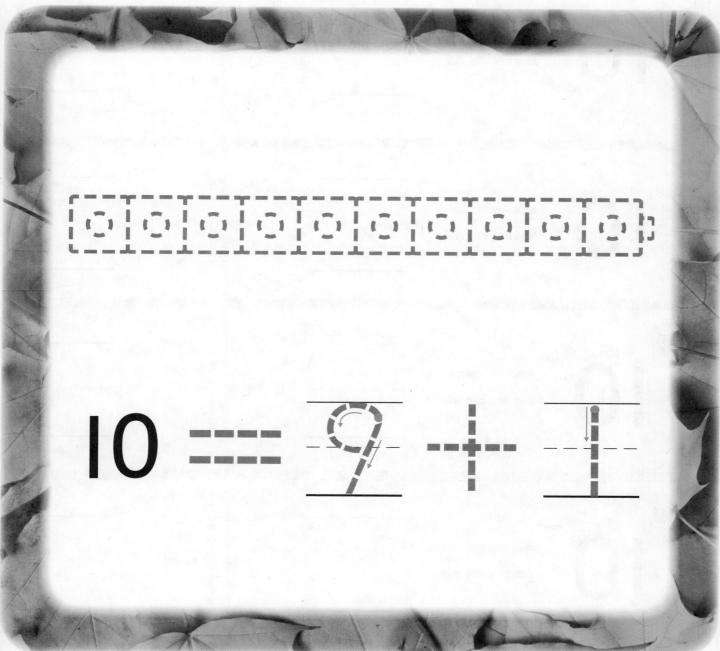

$$10 = 9 + 1$$

DIRECTIONS Listen to the word problem. Use two colors of cubes to make a cube train to match the addition sentence. Color the cube train to show your work. Trace the addition sentence.

Chapter 5 • Lesson 12   two hundred thirteen **213**

Share and Show

1. $10 = \underline{1} + \underline{9}$

2. $10 = \underline{\quad} + \underline{\quad}$

3. $10 = \underline{\quad} + \underline{\quad}$

4. $10 = \underline{\quad} + \underline{\quad}$

DIRECTIONS Listen to the word problem. Use two colors of cubes to build a cube train to show the number pairs that make 10. **1.** Trace the addition sentence to show one of the pairs. **2–4.** Complete the addition sentence to show a number pair for 10. Color the cube train to match the addition sentence in Exercise 4.

214 two hundred fourteen

Name _____

5 10 = = = _____ + _____

6 10 = = = _____ + _____

7 10 = = = _____ + _____

8 10 = = = _____ + _____

DIRECTIONS Listen to the word problem. Use two colors of cubes to build a cube train to show the number pairs that make 10. **5–8.** Complete the addition sentence to show a number pair for 10. Color the cube train to match the addition sentence in Exercise 8.

PROBLEM SOLVING REAL WORLD

❶

$$10 = \underline{\hspace{2cm}} + \underline{\hspace{2cm}}$$

②

$$10 = \underline{\hspace{2cm}} + \underline{\hspace{2cm}}$$

DIRECTIONS 1. There are ten children in the cafeteria. Ten of them are drinking water. How many children are drinking milk? Complete the addition sentence to show the number pair. **2.** Draw to show what you know about a different number pair for 10. Complete the addition sentence.

HOME ACTIVITY • Have your child tell you the number pairs for a set of ten objects. Have him or her tell the addition sentence to match one of the number pairs.

216 two hundred sixteen

FOR MORE PRACTICE:
Standards Practice Book, pp. P103–P104

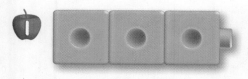 **Chapter 5 Review/Test**

Vocabulary

  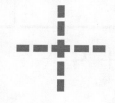   **Add**

Concepts and Skills

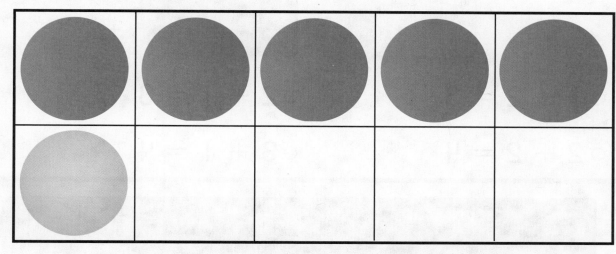

DIRECTIONS **1.** Trace to complete the addition sentence. (pp. 181–183)
2. How many are there of each color counter? Write the numbers and trace the symbol. Write the number to show how many in all.

Chapter 5   two hundred seventeen **217**

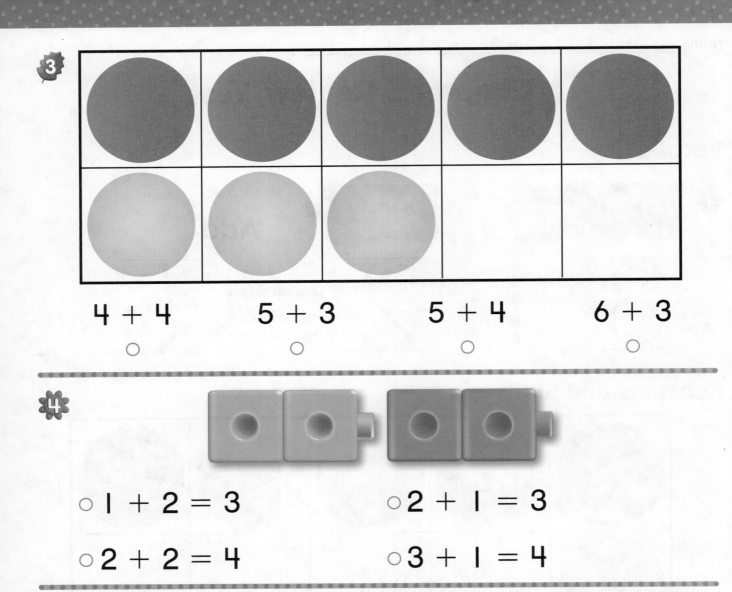

3

4 + 4   5 + 3   5 + 4   6 + 3
○   ○   ○   ○

4

○ 1 + 2 = 3   ○ 2 + 1 = 3

○ 2 + 2 = 4   ○ 3 + 1 = 4

5

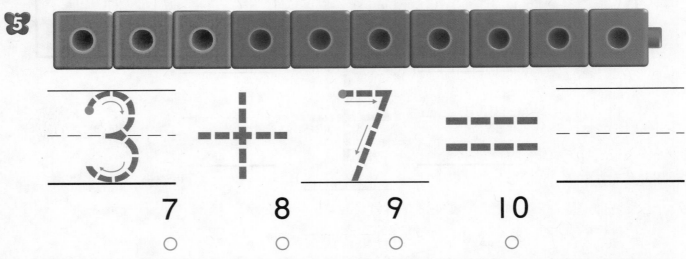

3 + 7 ===

7   8   9   10
○   ○   ○   ○

DIRECTIONS **3.** Mark under the addition that shows the sets put together. **4.** Mark beside the addition sentence that shows how many there are when the cubes are put together. **5.** How many red cubes do you see? How many blue cubes do see? How many cubes do you see in all? Mark under the number that would complete the addition sentence.

$$1 + 2 = \underline{}$$

1	2	3	4
○	○	○	○

$$3 + 5 = \underline{}$$

6	7	8	9
○	○	○	○

○ $8 = 2 + 6$ ○ $8 = 3 + 5$

○ $8 = 4 + 4$ ○ $8 = 5 + 3$

DIRECTIONS **6.** Tell an addition word problem about the sets. How many are there in all? Mark under the number that would complete the addition sentence. **7.** Tell an addition word problem about the sets. How many are there in all? Mark under the number that would complete the addition sentence. **8.** Mark beside the addition sentence that shows the number pair for the cube train.

Performance Task

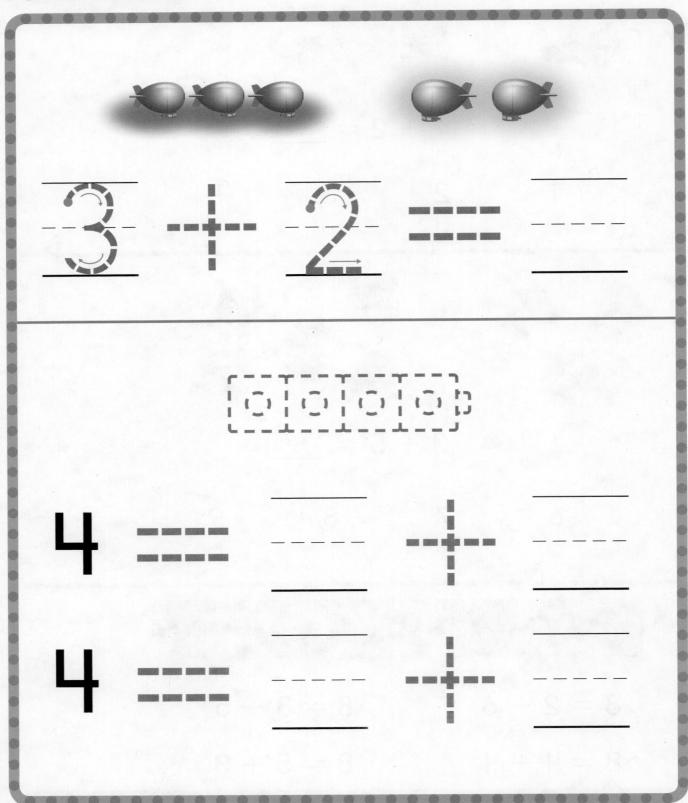

3 + 2 = ____

DIRECTIONS These tasks will assess the child's understanding of addition.

© Houghton Mifflin Harcourt Publishing Company

220 two hundred twenty

Chapter 6 Subtraction

Curious About Math with Curious George

Penguins are birds with black and white feathers.

- There are 4 penguins on the ice. One penguin jumps in the water. How many penguins are on the ice now?

Name _____

Show What You Know

Fewer

 1

 2

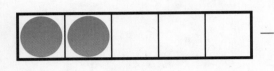

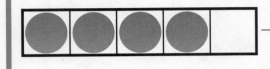

_____

- - - - - -

_____

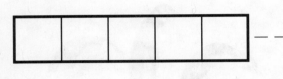

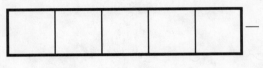

_____

- - - - - -

_____

Compare Numbers to 10

 3

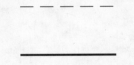

_____

- - - - - -

_____

_____

- - - - - -

_____

DIRECTIONS **1–2.** Count and tell how many. Draw a set with one fewer counter. Write how many in each set. **3.** Write the number of cubes in each set. Circle the number that is less than the other number.

FAMILY NOTE: This page checks your child's understanding of important skills needed for success in Chapter 6.

© Houghton Mifflin Harcourt Publishing Company

222 two hundred twenty-two

GO Online   Assessment Options
Soar to Success Math

Name _____

Vocabulary Builder

add

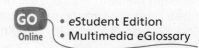

DIRECTIONS Add the set of bees and the set of butterflies. Write how many insects altogether.

GO
Online
• eStudent Edition
• Multimedia eGlossary

Game

Spin for More

Spin for More				
Player 1				
Player 2				

DIRECTIONS Play with a partner. Decide who goes first. Take turns spinning to get a number from each spinner. Use cubes to model a cube train with the number from the first spin. Say the number. Add the cubes from the second spin. Compare your number with your partner's. Mark an X on the table for the player who has the greater number. The first player to have five Xs wins the game.

MATERIALS two paper clips, connecting cubes

Name _____

Subtraction: Take From

Essential Question How can you show subtraction as taking from?

Listen and Draw  REAL WORLD

 **take away**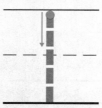

DIRECTIONS Listen to the subtraction word problem. Trace the number that shows how many children in all. Trace the number that shows how many children are leaving. Trace the number that shows how many children are left.

Chapter 6 • Lesson 1   two hundred twenty-five **225**

© Houghton Mifflin Harcourt Publishing Company

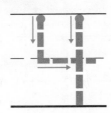 **take away** _____

_ _ _ _

_____

_ _ _ _

_____

DIRECTIONS 1. Listen to the subtraction word problem. Trace the number that shows how many children in all. Write the number that shows how many children are leaving. Write the number that shows how many children are left.

_____   _____

- - - - -   - - - - -

_____ **take away** _____

_____

- - - - -

_____

DIRECTIONS   2. Listen to the subtraction word problem Write the
number that shows how many children in all. Write the number that
shows how many children are leaving. Write the number that shows
how many children are left.

Chapter 6 • Lesson 1   two hundred twenty-seven   **227**

PROBLEM SOLVING REAL WORLD

1

_____

- - - - -   ▬▬▬   - - - -

_____   _____

2

_____

- - - - -

_____

DIRECTIONS   1. Blair has two marbles. His friend takes one marble from him. Draw to show the subtraction. Write the numbers and trace the symbol.   **2.** Write the number that shows how many marbles Blair has now.

HOME ACTIVITY • Show your child a set of four small objects. Have him or her tell how many objects there are. Take one of the objects from the set. Have him or her tell you how many objects there are now.

228 two hundred twenty-eight

FOR MORE PRACTICE:
Standards Practice Book, pp. P109–P110

Name _____

Subtraction: Take Apart

Essential Question How can you show subtraction as taking apart?

Listen and Draw

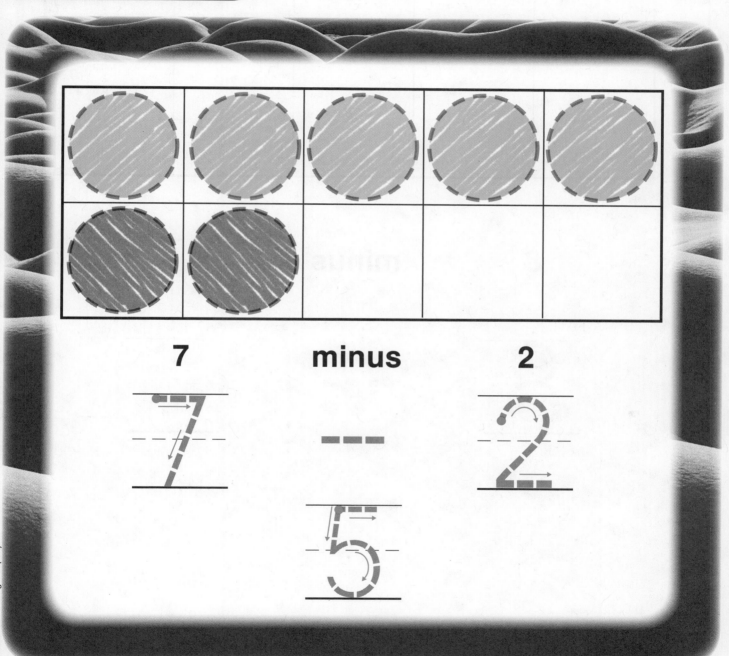

7   **minus**   **2**

DIRECTIONS Listen to the subtraction word problem. Place seven counters in the ten frame. Trace the counters. Trace the number that shows how many in all. Trace the number that shows how many are red. Trace the number that shows how many are yellow.

Chapter 6 • Lesson 2   two hundred twenty-nine **229**

Share and Show

8   **minus**   **1**

DIRECTIONS   **1.** Listen to the subtraction word problem. Place eight counters in the ten frame. Draw and color the counters. Trace the number that shows how many in all. Write the number that shows how many are yellow. Write the number that shows how many are red.

Name _____

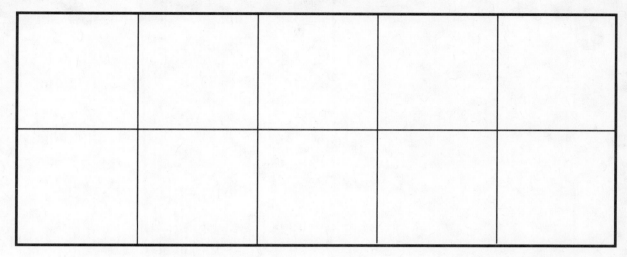

10   **minus**   **4**

_____   _____

– – – – – –   – – – – – –

_____   _____

_____

– – – – – –

_____

DIRECTIONS 2. Listen to the subtraction word problem. Place ten
counters in the ten frame. Draw and color the counters. Write the number
that shows how many in all. Write the number that shows how many are
red. Write the number that shows how many are yellow.

Chapter 6 • Lesson 2   two hundred thirty-one **231**

PROBLEM SOLVING REAL WORLD

1

_____

- - - - - - -   ▬ ▬ ▬   _____

_____   _____

2

_____

- - - - - - -

_____

DIRECTIONS **1.** Juli has nine apples. One apple is red. The rest of the apples are yellow. Draw the apples. Write the numbers and trace the symbol. **2.** Write the number that shows how many apples are yellow.

HOME ACTIVITY • Show your child a set of seven small objects. Now take away four objects. Have him or her tell a subtraction word problem about the objects.

FOR MORE PRACTICE:
Standards Practice Book, pp. P111–P112

Name _____

Problem Solving • Act Out Subtraction Problems

Essential Question How can you solve problems using the strategy *act it out*?

Unlock the Problem REAL WORLD

 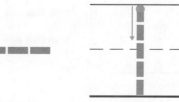  

3 − 1 = 2

DIRECTIONS Listen to and act out the subtraction word problem. Trace the subtraction sentence. How can you use subtraction to tell how many children are left?

Chapter 6 • Lesson 3

two hundred thirty-three **233**

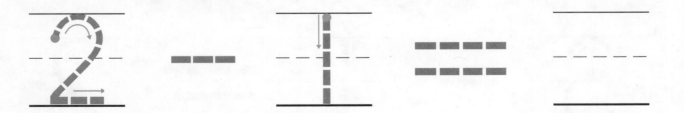

DIRECTIONS   I. Listen to and act out the subtraction word problem. Trace the numbers and the symbols. Write the number that shows how many children are left.

Name _____

Share and Show

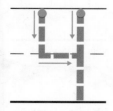  

DIRECTIONS   2. Listen to and act out the subtraction word problem. Trace the numbers and the symbols. Write the number that shows how many children are left.

Chapter 6 • Lesson 3   two hundred thirty-five   **235**

On Your Own

1

$$4 - 1 = \underline{\hspace{2cm}}$$

2

$$4 - 3 = \underline{\hspace{2cm}}$$

DIRECTIONS 1. Tell a subtraction word problem about the kittens. Trace the numbers and the symbols. Write the number that shows how many kittens are left. **2.** Draw to show what you know about the subtraction sentence. Write how many are left. Tell a friend a subtraction word problem to match.

HOME ACTIVITY • Tell your child a short subtraction word problem. Have him or her use objects to act out the word problem.

FOR MORE PRACTICE:
Standards Practice Book, pp. P113–P114

Name _____

Algebra • Model and Draw Subtraction Problems

Essential Question How can you use objects and drawings to solve subtraction word problems?

Listen and Draw

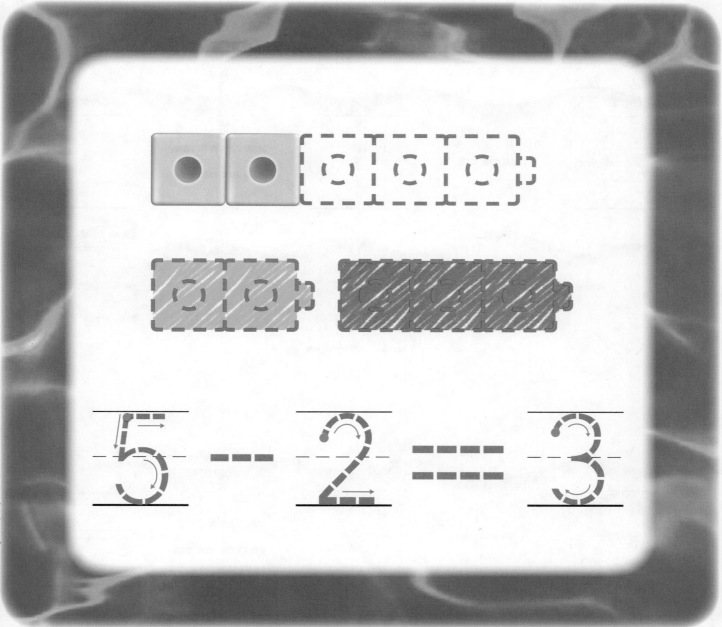

DIRECTIONS Model a five-cube train. Two cubes are yellow and the rest are red. Take apart the train to show how many cubes are red. Draw and color the cube trains. Write and trace to complete the subtraction sentence.

© Houghton Mifflin Harcourt Publishing Company

Chapter 6 • Lesson 4

Share and Show

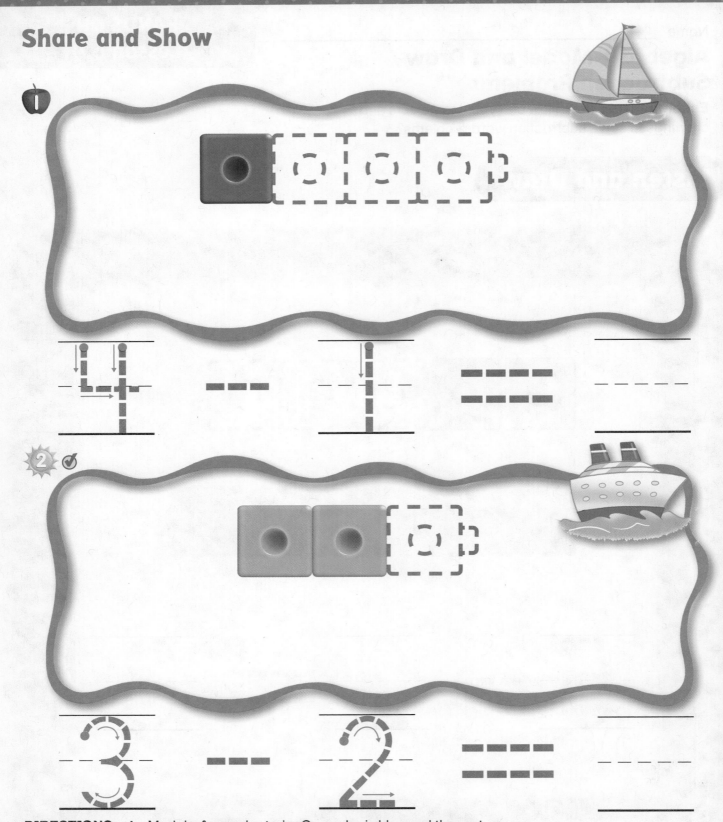

1. $4 - 1 = $ ___

2. $3 - 2 = $ ___

DIRECTIONS **1.** Model a four-cube train. One cube is blue and the rest are green. Take apart the train to show how many cubes are green. Draw and color the cube trains. Trace and write to complete the subtraction sentence. **2.** Model a three-cube train. Two cubes are orange and the rest are blue. Take apart the train to show how many cubes are blue. Draw and color the cube trains. Trace and write to complete the subtraction sentence.

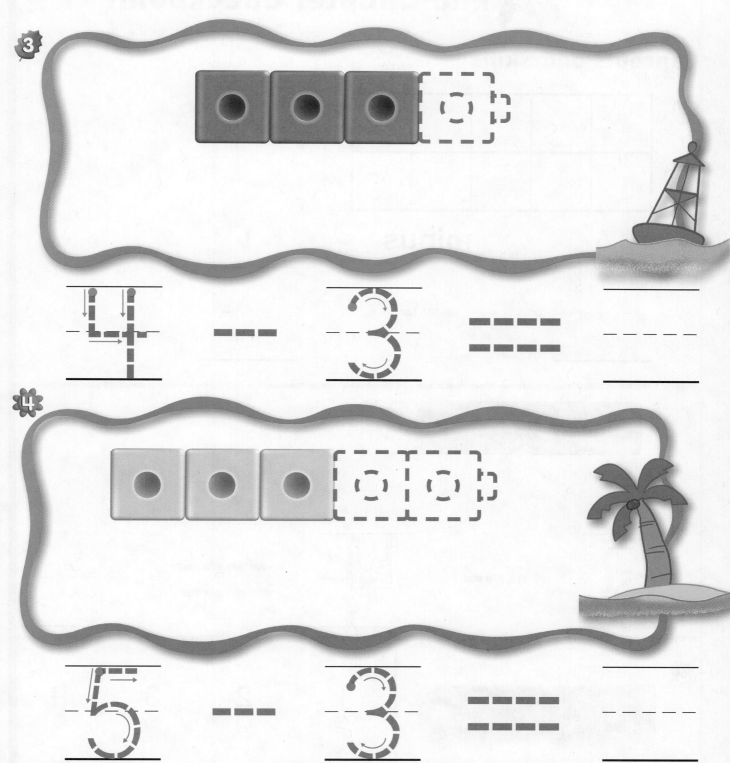

3

$$4 - 3 = \text{------}$$

4

$$5 - 3 = \text{------}$$

DIRECTIONS   3. Model a four-cube train. Three cubes are red and the rest are blue. Take apart the train to show how many cubes are blue. Draw and color the cube trains. Trace and write to complete the subtraction sentence. **4.** Model a five-cube train. Three cubes are yellow and the rest are green. Take apart the train to show how many cubes are green. Draw and color the cube trains. Trace and write to complete the subtraction sentence.

HOME ACTIVITY • Show your child two small objects. Take apart the set of objects. Have him or her tell a word problem to match the subtraction.

Chapter 6 • Lesson 4

FOR MORE PRACTICE:
Standards Practice Book, pp. P115–P116

two hundred thirty-nine **239**

Concepts and Skills

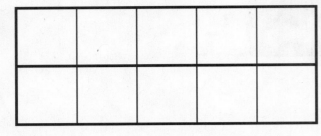

6   **minus**   **l**

_____   **- - -**   _____

- - - - -      - - - - -

_____   _____

  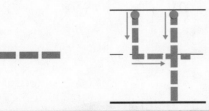

_____

5 **- -** **4** **═ ═** **- - - -**

_____

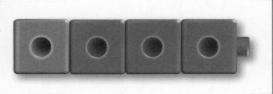

| **l** | **2** | **3** | **4** |
| ○ | ○ | ○ | ○ |

DIRECTIONS **I.** Kim has six counters. One is yellow and the others are red. Draw and color the counters. Write the number that shows how many in all. Write the number that shows how many are yellow. **2.** Model a five-cube train. Four cubes are blue and the rest are orange. Take apart the cube train to show how many are orange. Draw and color the cube trains. Trace and write to complete the subtraction sentence. **3.** Mark under the number that shows how many are red.

Name _____

Algebra • Write Subtraction Sentences

Essential Question How can you solve subtraction word problems and complete the equation?

Listen and Draw REAL WORLD

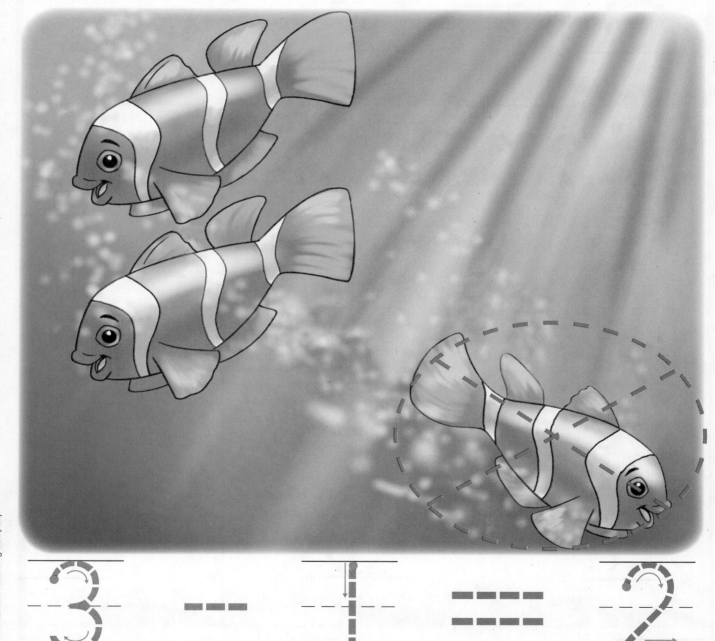

DIRECTIONS There are three fish. One fish swam away. Now there are two fish. Trace the circle and X to show the fish swimming away. Trace the subtraction sentence.

Chapter 6 • Lesson 5

two hundred forty-one **241**

Share and Show

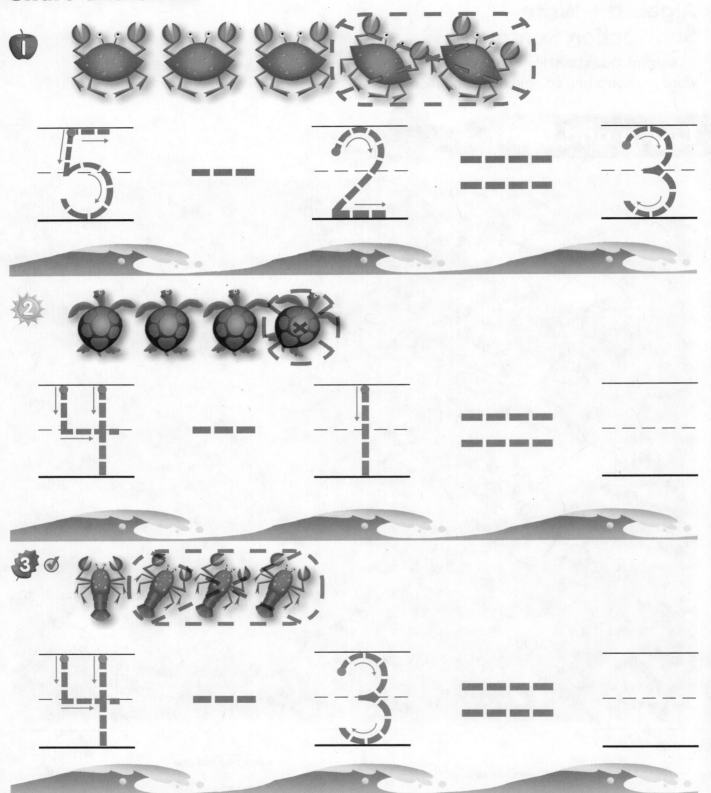

1

5 − 2 = 3

2

4 − 1 = ___

3 ✓

4 − 3 = ___

DIRECTIONS **1.** Listen to the subtraction word problem. Trace the circle and X to show how many are being taken from the set. Trace to complete the subtraction sentence. **2–3.** Listen to the subtraction word problem. Trace the circle and X to show how many are being taken from the set. Trace and write to complete the subtraction sentence.

Name _____

4

5   −   3   =   _ _ _ _

5

3   −   2   =   _ _ _ _

6

5   −   4   =   _ _ _ _

DIRECTIONS **4–6.** Listen to the subtraction word problem. Trace the circle and X to show how many are being taken from the set. Trace and write to complete the subtraction sentence.

© Houghton Mifflin Harcourt Publishing Company

PROBLEM SOLVING REAL WORLD

①

②

DIRECTIONS **1.** Kristen has four flowers. She gives her friend two flowers. Now how many flowers does Kristen have? Draw to solve the problem. Trace and write to complete the subtraction sentence. **2.** Tell a different subtraction word problem about the flowers. Draw to show how you solved the problem. Tell a friend about your drawing.

HOME ACTIVITY • Have your child draw a set of five or fewer balloons. Have him or her circle and mark an X on some balloons to show that they have popped. Then have your child tell a word problem to match the subtraction.

FOR MORE PRACTICE:
Standards Practice Book, pp. P117–P118

Name _____

Algebra • Write More Subtraction Sentences

Essential Question How can you solve subtraction word problems and complete the equation?

Listen and Draw REAL WORLD

DIRECTIONS There are six birds. One bird flies away. Trace the circle and X around that bird. How many birds are left? Trace the subtraction sentence.

Chapter 6 • Lesson 6

Share and Show

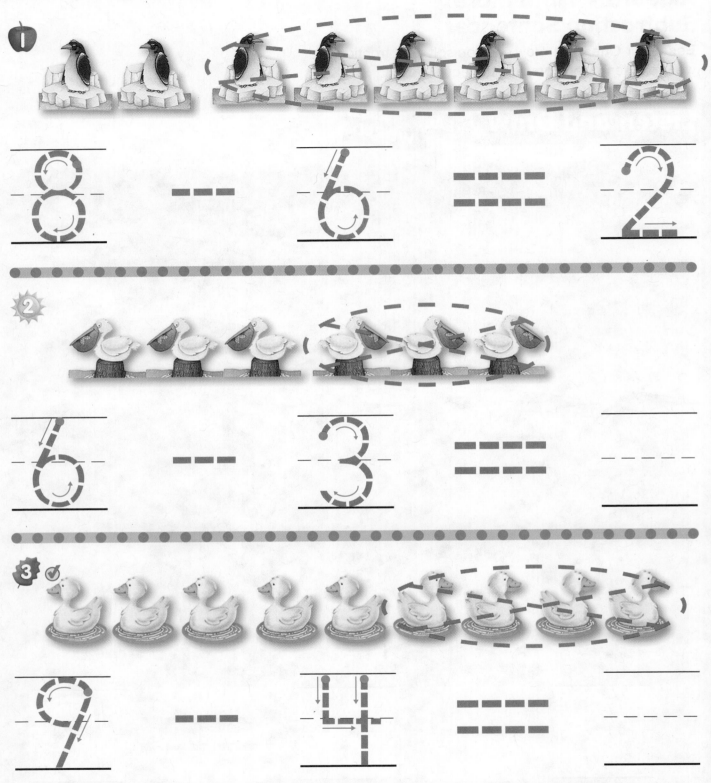

1   8 — 6 = 2

2   6 — 3 =

3   9 — 4 =

DIRECTIONS   **I.** Listen to the subtraction word problem. How many birds are taken from the set? Trace the circle and X. How many birds are left? Trace to complete the subtraction sentence.   **2–3.** Listen to the subtraction word problem. How many birds are taken from the set? Trace the circle and X. How many birds are left? Write the number to complete the subtraction sentence.

✿ 4

6 − 5 = ＝ ＿ ＿

❀ 5

9 − 6 = ＝ ＿ ＿

🌰 6

8 − 3 = ＝ ＿ ＿

DIRECTIONS **4–6.** Listen to the subtraction word problem. How many birds are taken from the set? Trace the circle and X. How many birds are left? Write the number to complete the subtraction sentence.

PROBLEM SOLVING

$$8 - 6 = \underline{}$$

DIRECTIONS Complete the subtraction sentence. Draw to show what you know about this subtraction sentence. Tell a friend about your drawing.

HOME ACTIVITY • Tell your child you have seven small objects. Tell him or her that you are taking two objects from the set. Ask him or her to tell you how many objects are left in the set.

FOR MORE PRACTICE:
Standards Practice Book, pp. P119–P120

Name _____

Algebra • Addition and Subtraction

Essential Question How can you solve word problems using addition and subtraction?

Listen and Draw

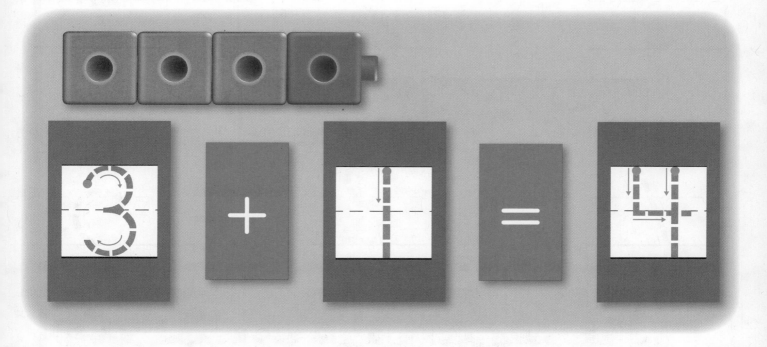

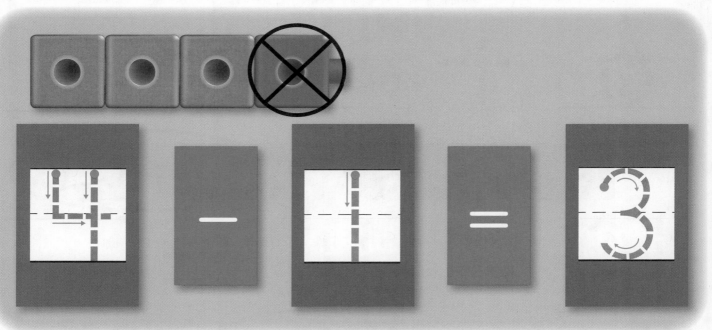

DIRECTIONS Listen to the addition and subtraction word problems. Use cubes and Number and Symbol Tiles as shown to match the word problems. Trace to complete the number sentences.

Share and Show

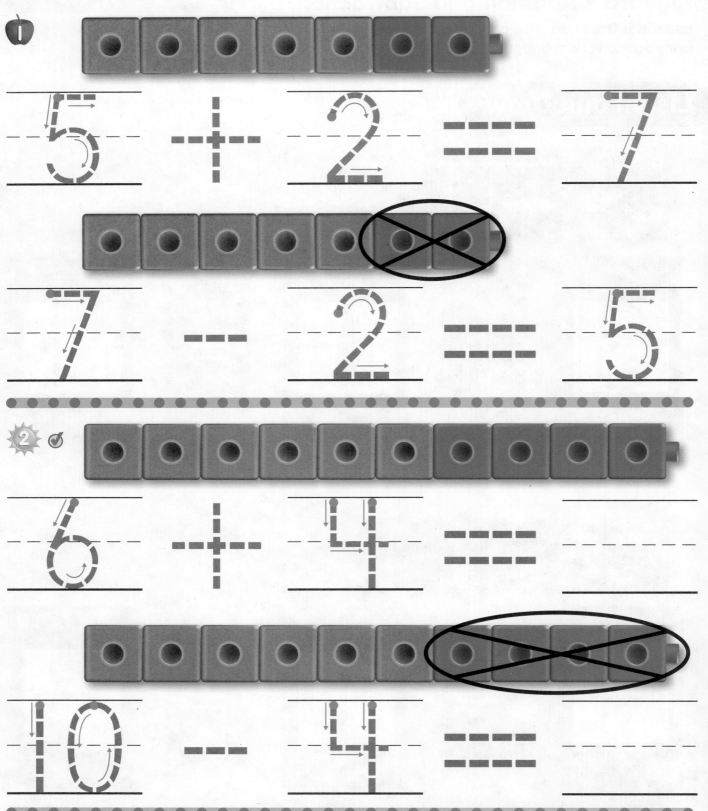

1. 5 + 2 = 7

7 - 2 = 5

2. ✓ 6 + 4 = ___

10 - 4 = ___

DIRECTIONS Tell addition and subtraction word problems. Use cubes to add and to subtract. **1.** Trace the number sentences. **2.** Trace and write to complete the number sentences.

3

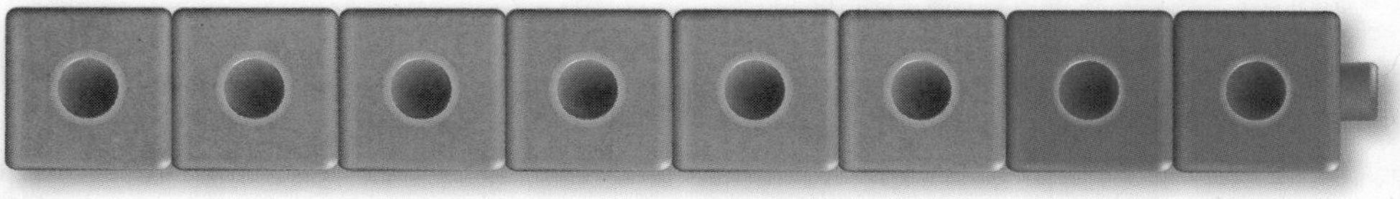

6 + 2 = = _ _ _

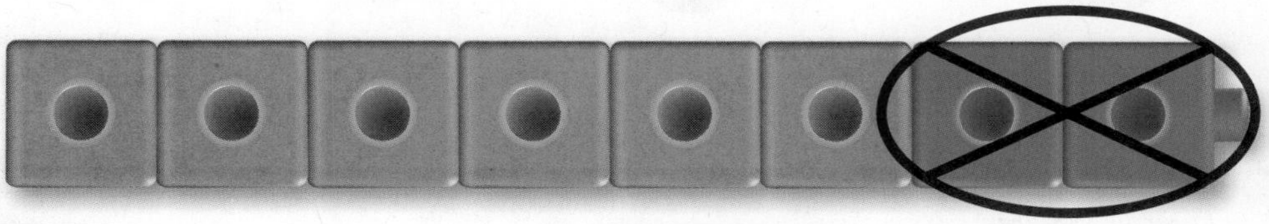

8 - 2 = = _ _ _

4

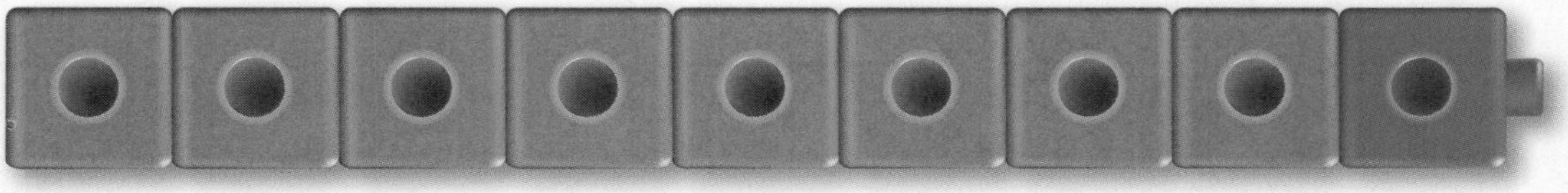

8 + 1 = = _ _ _

9 - 1 = = _ _ _

DIRECTIONS   3–4. Tell addition and subtraction word problems.
Use cubes to add and subtract. Trace and write to complete the
number sentences.

Chapter 6 • Lesson 7

PROBLEM SOLVING

$$6 + 3 = 9$$

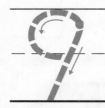

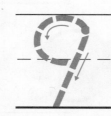

DIRECTIONS Look at the addition sentence at the top of the page. **1–2.** Tell a related subtraction word problem. Complete the subtraction sentence.

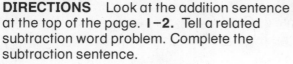

HOME ACTIVITY • Ask your child to use objects to model a simple addition problem. Then have him or her explain how to make it into a subtraction problem.

FOR MORE PRACTICE:
Standards Practice Book, pp. P121–P122

Chapter 6 Review/Test

Vocabulary

subtract

_____

- - - - - - - - - - - - - - -

_____

Concepts and Skills

 9 − 3 = = = ___

5 − 2 = = = ___

DIRECTIONS 1. Write the number to show how many are left. (pp. 233–236) 2. Model a nine-cube train. Three cubes are yellow and the rest are blue. Take apart the cube train to show how many are blue. Draw and color the cube trains. Trace and write to complete the subtraction sentence. 3. There are five birds. Two birds are taken from the set. How many birds are left? Trace and write to complete the subtraction sentence.

4

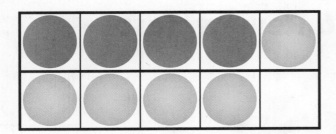

○ $9 - 5$ ○ $9 - 1$

○ $8 - 3$ ○ $5 - 4$

5

$8 - 1 =$ ____

1	5	7	8
○	○	○	○

6

$7 - 3 =$ ____

2	3	4	5
○	○	○	○

DIRECTIONS **4.** Mark beside the subtraction that shows how many counters are red. **5.** Mark under the number that shows how many cubes are left. **6.** Mark under the number that shows how many birds are left.

$$8 - 2 = \underline{\quad}$$

5	6	7	8
○	○	○	○

○ 2 + 1 = 3   ○ 3 + 1 = 4

○ 3 + 2 = 5   ○ 4 + 2 = 6

○ 3 − 2 = 1   ○ 4 − 1 = 3

○ 5 − 1 = 4   ○ 5 − 2 = 3

DIRECTIONS 7. Mark under the number that shows how many birds are left. **8.** Mark beside the addition sentence that matches the picture. **9.** Mark beside the subtraction sentence that matches the picture.

Performance Task

_____   __ __ __ __   _____   ▀▀▀ ▀▀▀ ▀▀▀   _____
- - - - -   ▀▀▀ ▀ ▀▀▀   - - - - -   ▀▀▀ ▀▀▀ ▀▀▀   - - - - -
_____   __ __ __ __   _____

_____   __ __ __ __   _____   ▀▀▀ ▀▀▀ ▀▀▀   _____
- - - - -   ▀▀ ▀▀ ▀▀   - - - - -   ▀▀▀ ▀▀▀ ▀▀▀   - - - - -
_____   __ __ __ __   _____

PERFORMANCE TASK   This task will assess the child's understanding
of addition and subtraction.

Chapter 7

Represent, Count, Read, and Write 11 to 19

Curious About Math with Curious George

Shells come in many colors and patterns.

- Is the number of shells greater than or less than 10?

Name _____

Draw Objects to 10

 1

10

 2

9

Write Numbers to 10

 3

_____

- - - - - - - - - - - -

_____

 4

_____

- - - - - - - - - - - -

_____

5

_____

- - - - - - - - - - - -

_____

6

_____

- - - - - - - - - - - -

_____

DIRECTIONS **1.** Draw 10 oranges. **2.** Draw 9 apples. **3–6.** Count and tell how many. Write the number.

FAMILY NOTE: This page checks your child's understanding of important skills needed for success in Chapter 7.

 **GO** Online **Assessment Options** **Soar to Success Math**

Name _____

Vocabulary Builder

FINISH

three

four

one 1

two

2

3

4

five

5

eight

8

seven

7

six

6

nine

9

ten

10

CABBAGE PARK RUN

DIRECTIONS Circle the number word that is greater than nine.

© Houghton Mifflin Harcourt Publishing Company

Game Sweet and Sour Path

DIRECTIONS Play with a partner. Place game markers on START. Take turns. Toss the number cube. Move that number of spaces. If a player lands on a lemon, the player reads the number and moves back that many spaces. If a player lands on a strawberry, the player reads the number and moves forward that many spaces. The first player to reach END wins.

MATERIALS two game markers, number cube (1–6)

Name _____

Model and Count 11 and 12

Essential Question How can you use objects to show
11 and 12 as ten ones and some more ones?

Listen and Draw

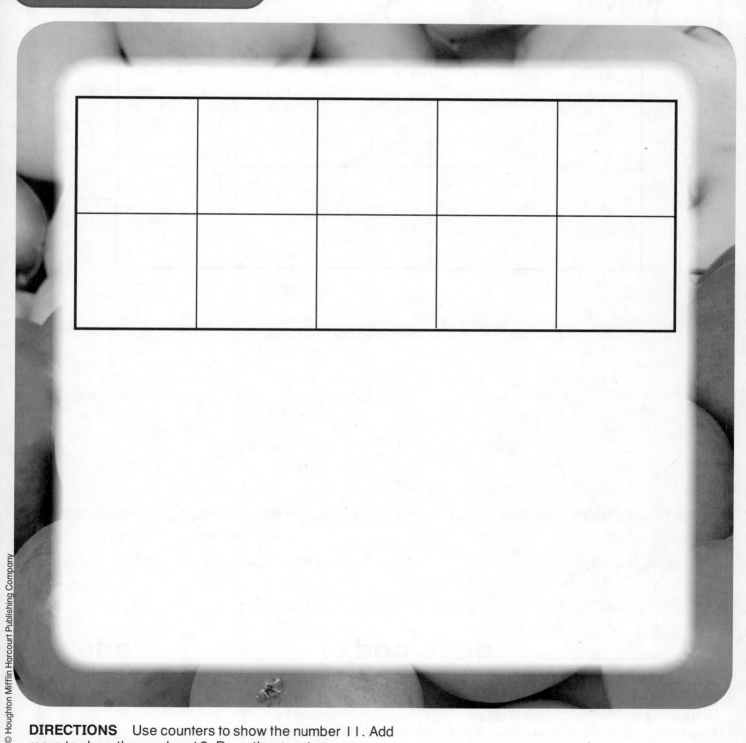

DIRECTIONS Use counters to show the number 11. Add
more to show the number 12. Draw the counters.

Chapter 7 • Lesson 1

two hundred sixty-one **261**

Share and Show

1

| |

eleven

2 ✓

3

10

_____ **ones and** _____ **one**

DIRECTIONS 1. Count and tell how many. Trace the number. **2.** Use counters to show the number 11. Draw the counters. **3.** Look at the counters you drew. How many ones are in the ten frame? Trace the number. How many more ones are there? Write the number.

Name _____

4 **12**
twelve

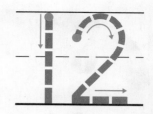

5

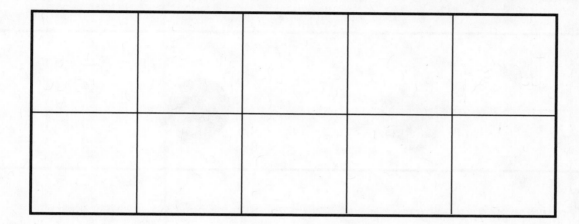

6

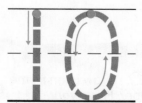   _____

_____ **ones and** _____ **ones**

DIRECTIONS **4.** Count and tell how many. Trace the number. **5.** Use counters to show the number 12. Draw the counters. **6.** Look at the counters you drew. How many ones are in the ten frame? Trace the number. How many more ones are there? Write the number.

Chapter 7 · Lesson 1   two hundred sixty-three **263**

PROBLEM SOLVING REAL WORLD

1

2

3

DIRECTIONS **1.** Start with the blue bead on the left. Circle to show 11 beads on the bead string. **2.** Are there more blue beads or more yellow beads in those 11 beads? Circle the color bead that has more. **3.** Draw a set of 11 objects. If you circle 10 of the objects, how many more objects are there? Complete the addition sentence to match.

HOME ACTIVITY • Draw a ten frame on a sheet of paper. Have your child use small objects such as buttons, pennies, or dried beans to show the numbers 11 and 12.

FOR MORE PRACTICE:
Standards Practice Book, pp. P127–P128

Name _____

Read and Write 11 and 12

Essential Question How can you read and write 11 and 12 with words and numbers?

Listen and Draw

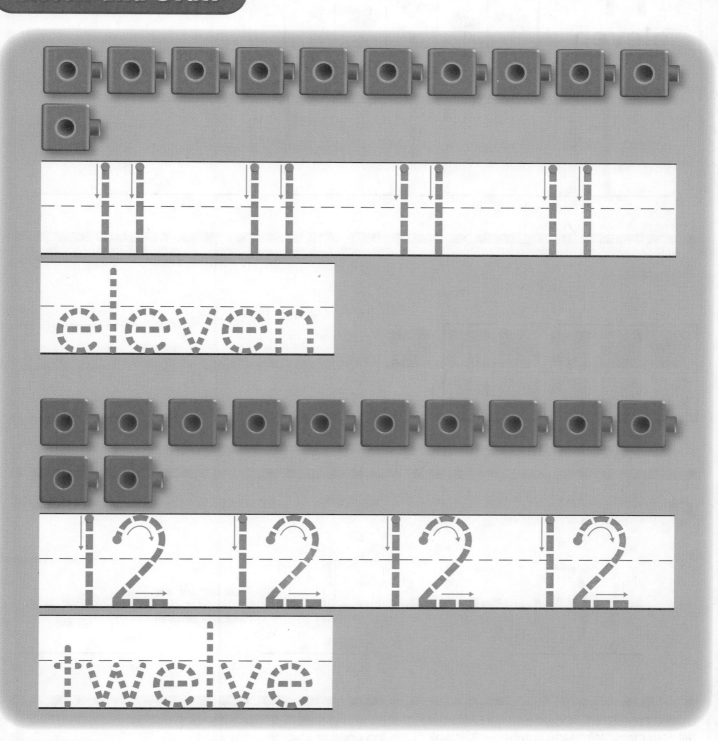

DIRECTIONS Count and tell how many. Trace the numbers and the words.

Share and Show

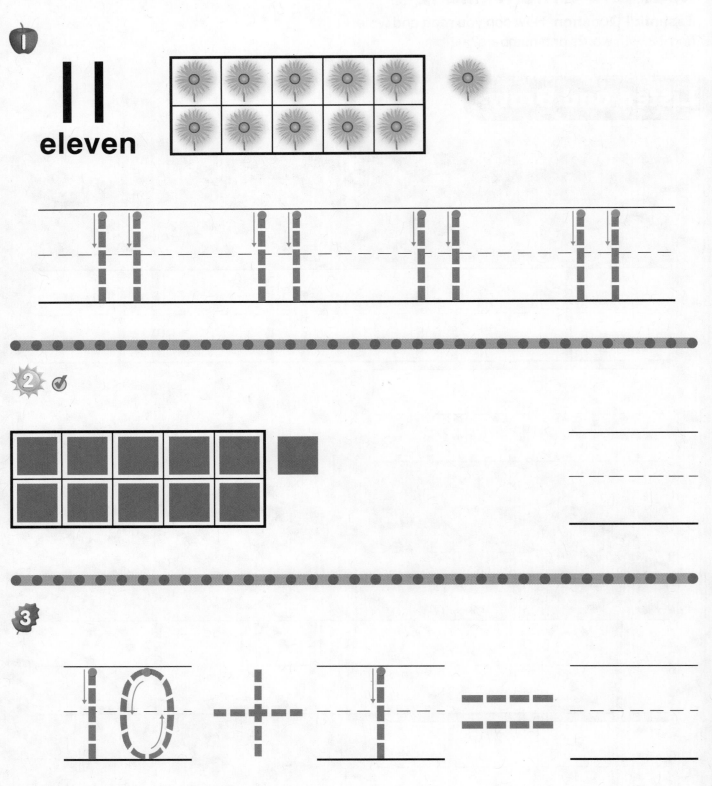

1

11
eleven

2 ✓

3

DIRECTIONS **1.** Count and tell how many. Trace the numbers. **2.** Count and tell how many. Write the number. **3.** Look at the ten ones and some more ones in Exercise 2. Complete the addition sentence to match.

4

12
twelve

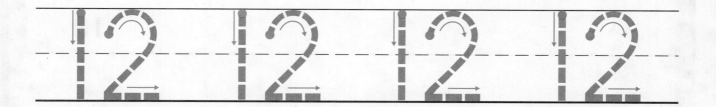

5

_____

- - - - - - - - -

_____

6

DIRECTIONS 4. Count and tell how many. Trace the numbers.   **5.** Count and tell how many. Write the number.   **6.** Look at the ten ones and some more ones in Exercise 5. Complete the addition sentence to match.

PROBLEM SOLVING REAL WORLD

①

11

12

13

②

12 === _____ + _____
    _____   _____

DIRECTIONS **1.** Circle a number. Draw more flowers to show that number. **2.** Draw a set of 12 objects. If you circle 10 of the objects, how many more objects are there? Complete the addition sentence to match.

HOME ACTIVITY • Ask your child to count and write the number for a set of 11 or 12 objects, such as macaroni pieces or buttons.

FOR MORE PRACTICE:
Standards Practice Book, pp. P129–P130

Name _____

Model and Count 13 and 14

Essential Question How can you use objects to show 13 and 14 as ten ones and some more ones?

Listen and Draw

DIRECTIONS Use counters to show the number 13. Add more to show the number 14. Draw the counters.

Chapter 7 • Lesson 3

1

13
thirteen

2 ✓

3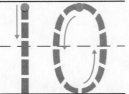

10

_____

- - - - - -

ones and _____ **ones**

DIRECTIONS 1. Count and tell how many. Trace the number. 2. Use counters to show the number 13. Draw the counters. 3. Look at the counters you drew. How many ones are in the ten frame? Trace the number. How many more ones are there? Write the number.

Name _____

4

14
fourteen

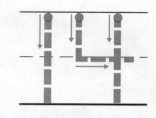

5

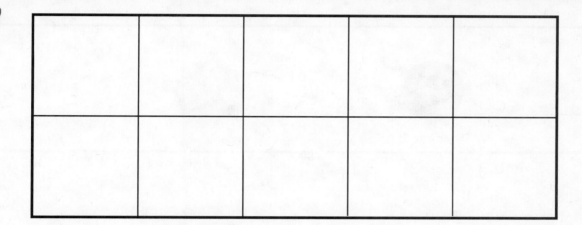

6

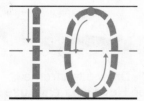

_____

ones and _____ **ones**

DIRECTIONS **4.** Count and tell how many. Trace the number. **5.** Use counters to show the number 14. Draw the counters. **6.** Look at the counters you drew. How many ones are in the ten frame? Trace the number. How many more ones are there? Write the number.

Chapter 7 • Lesson 3   two hundred seventy-one **271**

PROBLEM SOLVING REAL WORLD

1

2

3

13 = ___ ___ + ___

DIRECTIONS **1.** Start with the blue bead on the left. Circle to show 13 beads on the bead string. **2.** Are there more blue beads or more yellow beads in those 13 beads? Circle the color bead that has more. **3.** Draw a set of 13 objects. If you circle 10 of the objects, how many more objects are there? Complete the addition sentence to match.

HOME ACTIVITY • Draw a ten frame on a sheet of paper. Have your child use small objects such as buttons, pennies, or dried beans to show the numbers 13 and 14.

FOR MORE PRACTICE:
Standards Practice Book, pp. P131–P132

Name _____

Read and Write 13 and 14

Essential Question How can you read and write 13 and 14 with words and numbers?

Listen and Draw

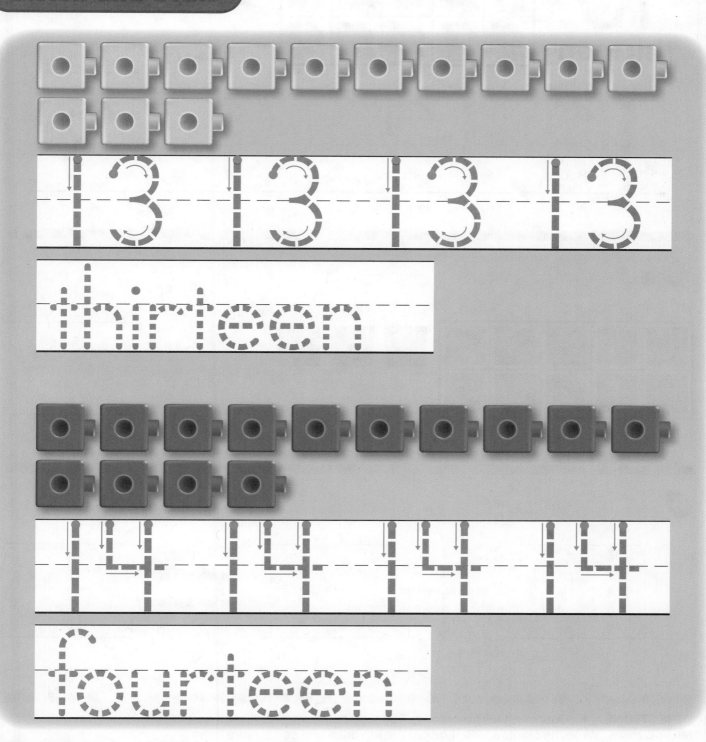

DIRECTIONS Count and tell how many. Trace the numbers and the words.

Share and Show

13
thirteen

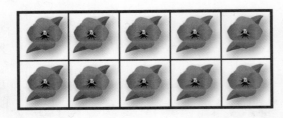

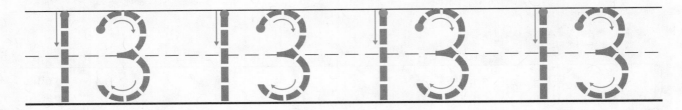

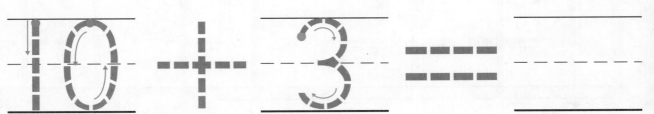

DIRECTIONS 1. Count and tell how many. Trace the numbers. 2. Count and tell how many. Write the number. 3. Look at the ten ones and some more ones in Exercise 2. Complete the addition sentence to match.

Name _____

fourteen

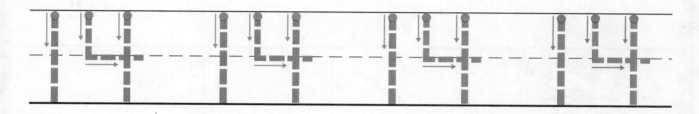

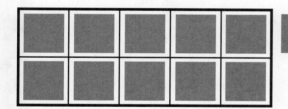

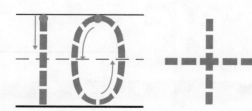 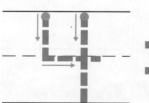 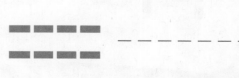

DIRECTIONS **4.** Count and tell how many. Trace the numbers. **5.** Count and tell how many. Write the number. **6.** Look at the ten ones and some more ones in Exercise 5. Complete the addition sentence to match.

PROBLEM SOLVING REAL WORLD

1

12

13

14

2

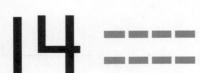 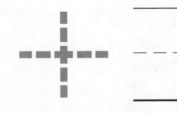

DIRECTIONS **1.** Circle a number. Draw more flowers to show that number. **2.** Draw a set of 14 objects. If you circle 10 of the objects, how many more objects are there? Complete the addition sentence to match.

HOME ACTIVITY • Ask your child to count and write the number for a set of 13 or 14 objects, such as macaroni pieces or buttons.

276 two hundred seventy-six

Name _____

Model, Count, and Write 15

Essential Question How can you use objects to show 15 as ten ones and some more ones and show 15 as a number?

Listen and Draw

DIRECTIONS Use counters to show the number 15. Draw the counters. Tell a friend about the counters.

Chapter 7 • Lesson 5

Share and Show

 1 **15**
fifteen

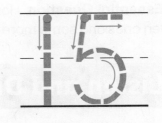

2 ✓

<table>
<tr><td></td><td></td><td></td><td></td><td></td></tr>
<tr><td></td><td></td><td></td><td></td><td></td></tr>
</table>

3 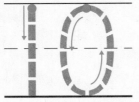   _____   ones and   _____   ones

---

DIRECTIONS   **1.** Count and tell how many. Trace the number.   **2.** Use counters to show the number 15. Draw the counters.   **3.** Look at the counters you drew. How many ones are in the ten frame? Trace the number. How many more ones? Write the number.

Name _____

15
fifteen

15 15 15 15

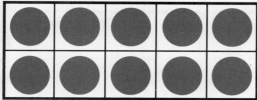

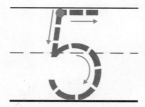

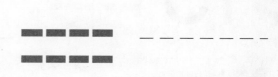

_____

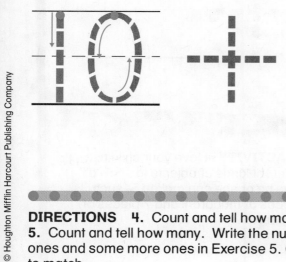

10 + 5 == _____

DIRECTIONS **4.** Count and tell how many. Trace the numbers.
5. Count and tell how many. Write the number. **6.** Look at the ten
ones and some more ones in Exercise 5. Complete the addition sentence
to match.

PROBLEM SOLVING REAL WORLD

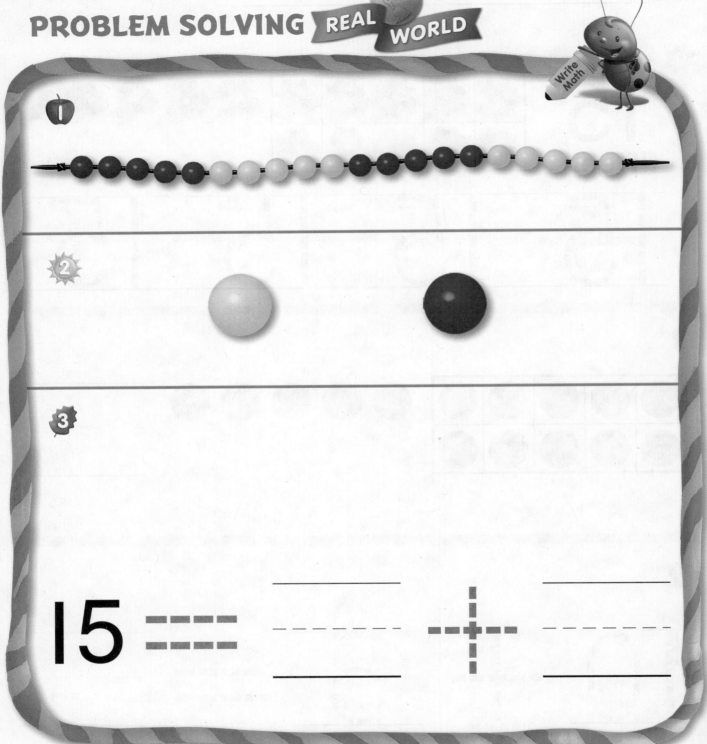

$$15 = \text{\_\_\_\_\_} + \text{\_\_\_\_\_}$$

DIRECTIONS **1.** Start with the blue bead on the left. Circle to show 15 beads on the bead string. **2.** Are there more blue beads or more yellow beads in those 15 beads? Circle the color bead that has more. **3.** Draw a set of 15 objects. If you circle 10 of the objects, how many more objects are there? Complete the addition sentence to match.

HOME ACTIVITY • Have your child use two different kinds of objects to show all the ways he or she can make 15, such as 8 pieces of macaroni and 7 pieces of bowtie pasta.

280 two hundred eighty

FOR MORE PRACTICE: Standards Practice Book, pp. P135–P136

Name _____

Problem Solving • Use Numbers to 15

Essential Question How can you solve problems using the strategy *draw a picture*?

🔑 Unlock the Problem REAL WORLD

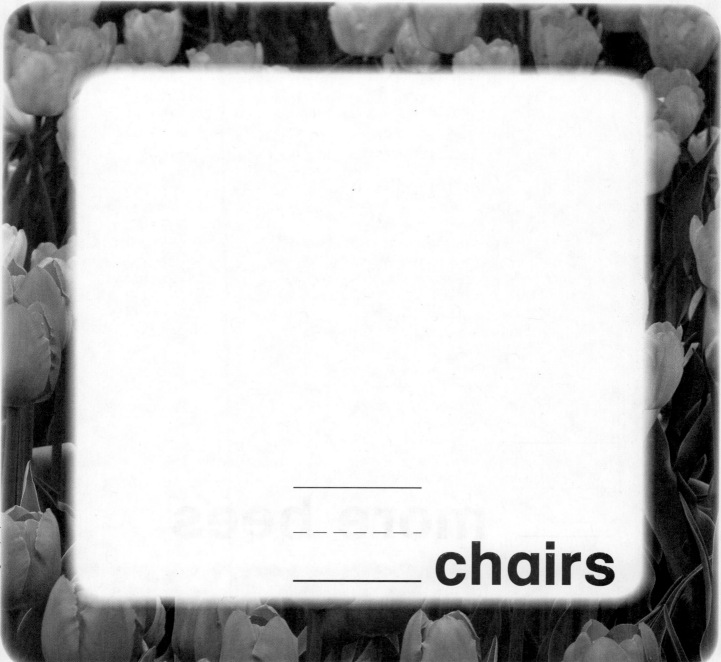

_____
_ _ _ _ _ _ _
_____ **chairs**

DIRECTIONS There are 14 children sitting on chairs. There is one chair with no child on it. How many chairs are there? Draw to show how you solved the problem.

Chapter 7 • Lesson 6

two hundred eighty-one **281**

Try Another Problem

I

____

_ _ _ _ _

_______ **more bees**

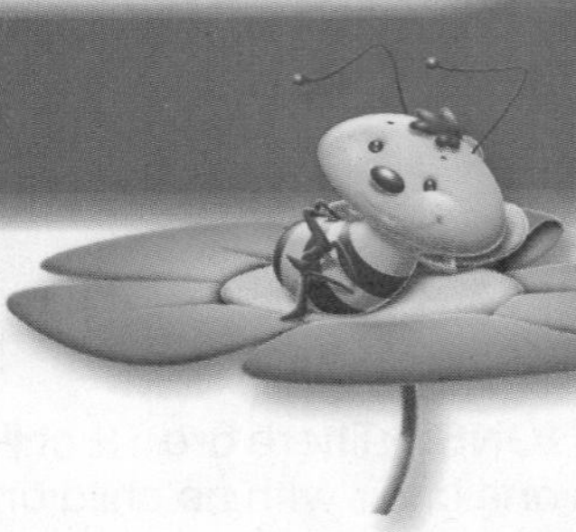

DIRECTIONS 1. There are 15 flowers. Ten flowers have 1 bee on them. How many more bees would you need to have one bee on each flower? Draw to solve the problem. Write how many more bees.

Name _____

Share and Show

___
_ _ _ _ _ _
_____ **boys**

© Houghton Mifflin Harcourt Publishing Company

DIRECTIONS 2. There are 15 children in Miss Sully's class. They sit in rows of 5. There are 3 boys and 2 girls in each row. How many boys are in the class? Draw to solve the problem.

HOME ACTIVITY • Draw a ten frame on a sheet of paper. Have your child use small objects such as buttons, pennies, or dried beans to show the number 15.

Concepts and Skills

 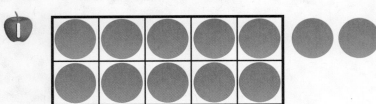

_____
- - - - - - - - -
_____

2

14 === _____ -+- - - - - - - - -
                - - - - - - - -   - - - - - - - -
                _____   _____

3 _____
- - - - - - - -
_____

4 _____
- - - - - - - -
_____

5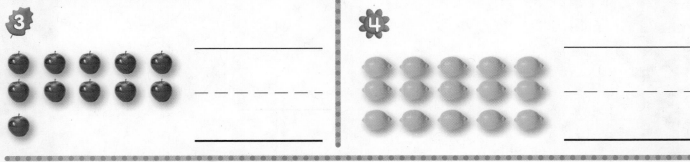

| 12 | 13 | 14 | 15 |
| ○ | ○ | ○ | ○ |

DIRECTIONS **1.** Count and tell how many. Write the number. **2.** Draw a set of 14 objects. If you circle 10 of the objects, how many more objects are there? Complete the addition sentence to match. **3–4.** Count and tell how many. Write the number. **5.** Mark under the number that shows how many flowers.

Name _____

Model and Count 16 and 17

Essential Question How can you use objects to
show 16 and 17 as ten ones and some more ones?

Listen and Draw

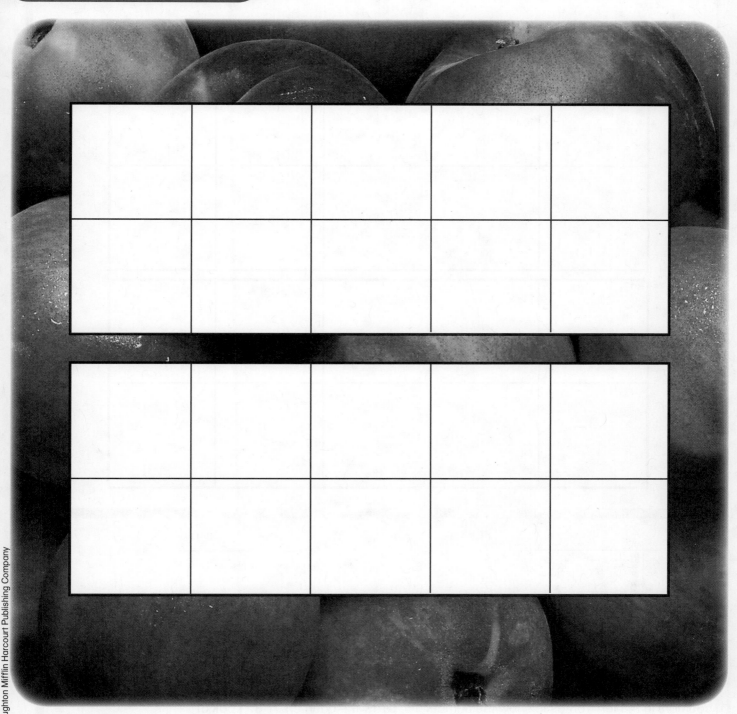

DIRECTIONS Use counters to show the number 16.
Add more to show the number 17. Draw the counters.

Chapter 7 • Lesson 7
two hundred eighty-five **285**

Share and Show

1  16
sixteen

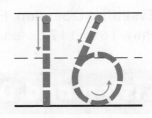

2 ✓

(two blank ten frames)

3

(traced number 10)

_____ ones and _____ ones

DIRECTIONS 1. Count and tell how many. Trace the number. **2.** Place counters in the ten frames to show the number 16. Draw the counters. **3.** Look at the counters you drew in the ten frames. How many ones are in the top ten frame? Trace the number. How many ones are in the bottom ten frame? Write the number.

4

17
seventeen

5

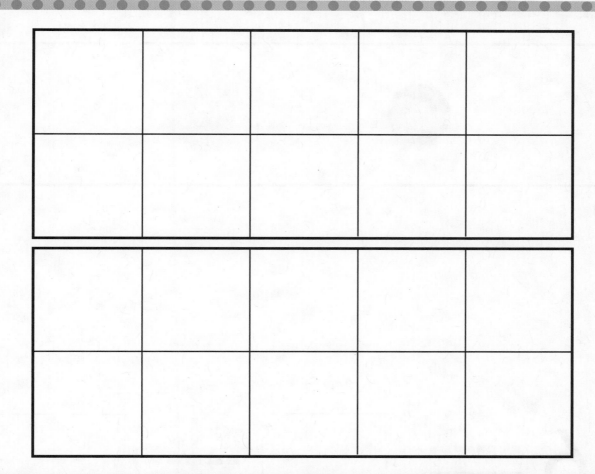

6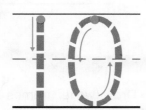

_____   _____

ones and _____ **ones**

DIRECTIONS **4.** Count and tell how many. Trace the number. **5.** Place counters in the ten frames to show the number 17. Draw the counters. **6.** Look at the counters you drew in the ten frames. How many ones are in the top ten frame? Trace the number. How many ones are in the bottom ten frame? Write the number.

Chapter 7 • Lesson 7

PROBLEM SOLVING

16 = === ____ + ____

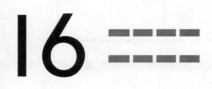

DIRECTIONS **1.** Start with the blue bead on the left. Circle to show 16 beads on the bead string. **2.** Are there more blue beads or more yellow beads in those 16 beads? Circle the color bead that has more. **3.** Draw a set of 16 objects. If you circle 10 of the objects, how many more objects are there? Complete the addition sentence to match.

HOME ACTIVITY • Draw two ten frames on a sheet of paper. Have your child use small objects such as buttons, pennies, or dried beans to show the numbers 16 and 17.

FOR MORE PRACTICE:
Standards Practice Book, pp. P139–P140

Name _____

Read and Write 16 and 17

Essential Question How can you read and write
16 and 17 with words and numbers?

Listen and Draw

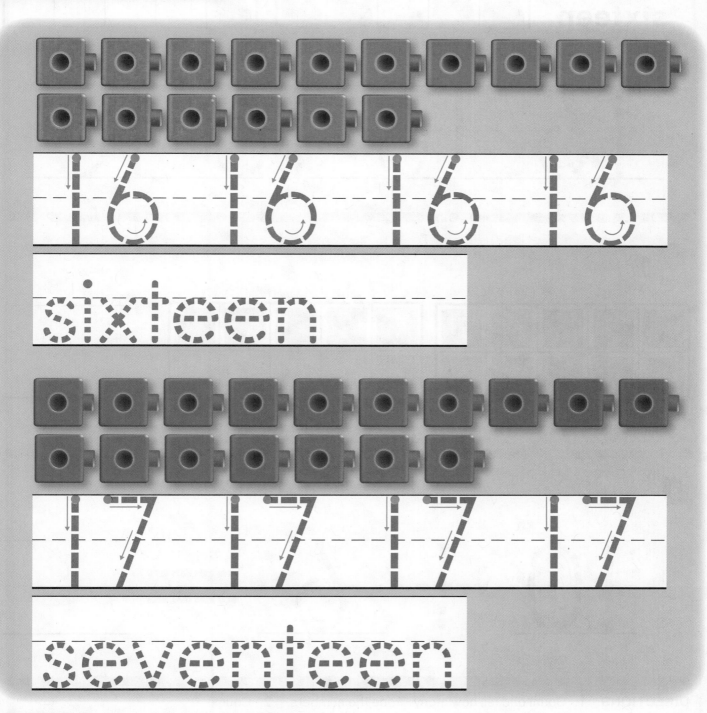

16 16 16 16

sixteen

17 17 17 17 17

seventeen

DIRECTIONS Count and tell how many. Trace the
numbers and the words.

Share and Show

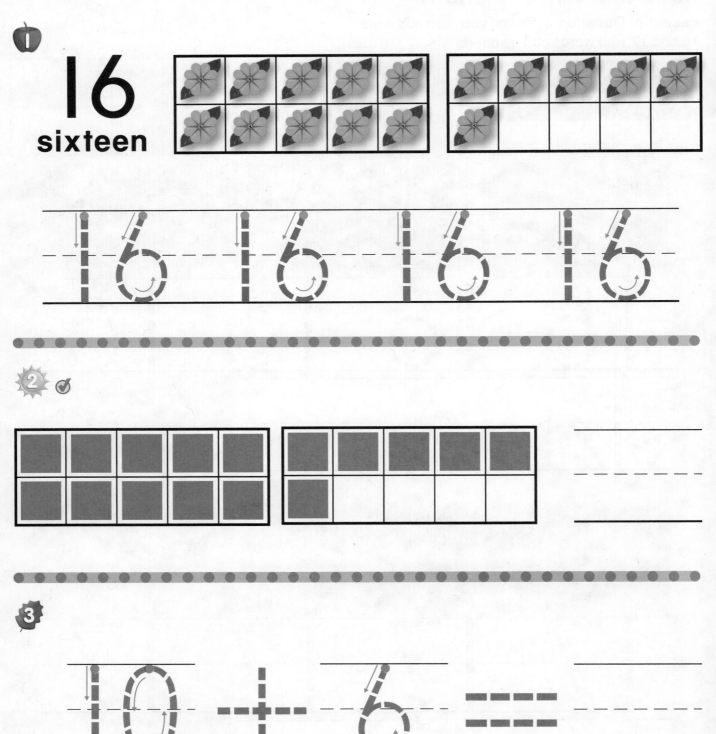

16
sixteen

2

3

DIRECTIONS **1.** Count and tell how many. Trace the numbers. **2.** Count and tell how many. Write the number. **3.** Look at the ten frames in Exercise 2. Complete the addition sentence to match.

290 two hundred ninety

4

17
seventeen

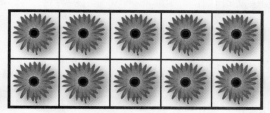

5

6

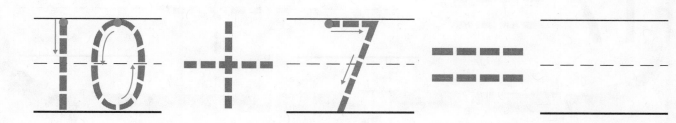

DIRECTIONS **4.** Count and tell how many. Trace the numbers. **5.** Count and tell how many. Write the number. **6.** Look at the ten frames in Exercise 5. Complete the addition sentence to match.

PROBLEM SOLVING REAL WORLD

1

17

18

19

2

17 === _____ **+** _____

DIRECTIONS 1. Circle a number. Draw more flowers to show that number. **2.** Draw a set of 17 objects. If you circle 10 of the objects, how many more objects are there? Complete the addition sentence to match.

HOME ACTIVITY • Ask your child to count and write the number for a set of 16 or 17 objects, such as macaroni pieces or buttons.

FOR MORE PRACTICE:
Standards Practice Book, pp. P141–P142

Name _____

Model and Count 18 and 19

Essential Question How can you use objects to show 18 and 19 as ten ones and some more ones?

Listen and Draw

DIRECTIONS Use counters to show the number 18. Add more to show the number 19. Draw the counters.

Share and Show

18
eighteen

 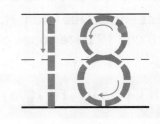

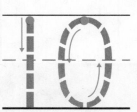 **ones and** _____ **ones**

DIRECTIONS **1.** Count and tell how many. Trace the number. **2.** Place counters in the ten frames to show the number 18. Draw the counters. **3.** Look at the counters you drew in the ten frames. How many ones are in the top ten frame? Trace the number. How many ones are in the bottom ten frame? Write the number.

4 # 19
nineteen

5

6

ones and _____ **ones**

DIRECTIONS 4. Count and tell how many. Trace the number. 5. Place counters in the ten frames to show the number 19. Draw the counters. 6. Look at the counters you drew in the ten frames. How many ones are in the top ten frame? Trace the number. How many ones are in the bottom ten frame? Write the number.

PROBLEM SOLVING REAL WORLD

$$18 = \rule{2cm}{0.4pt} + \rule{2cm}{0.4pt}$$

DIRECTIONS **1.** Start with the blue bead on the left. Circle to show 18 beads on the bead string. **2.** Are there more blue beads or more yellow beads in those 18 beads? Circle the color bead that has more. **3.** Draw a set of 18 objects. If you circle 10 of the objects, how many more objects are there? Complete the addition sentence to match.

HOME ACTIVITY • Draw two ten frames on a sheet of paper. Have your child use small objects such as buttons, pennies, or dried beans to model the numbers 18 and 19.

296 two hundred ninety-six

FOR MORE PRACTICE:
Standards Practice Book, pp. P143–P144

Name _____

Read and Write 18 and 19

Essential Question How can you read and write 18 and 19 with words and numbers?

Listen and Draw

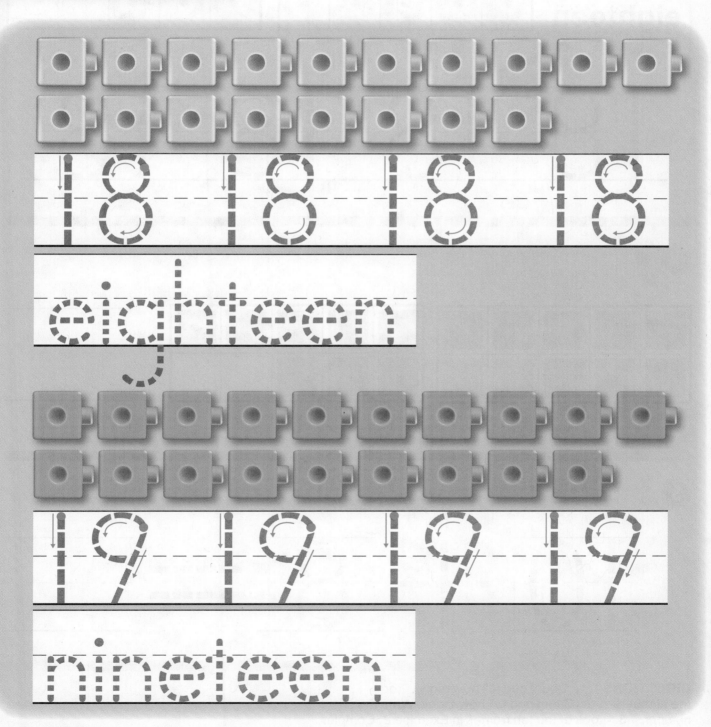

DIRECTIONS Count and tell how many. Trace the numbers and the words.

Share and Show

1

18
eighteen

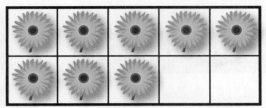

2 ✓

_____

- - - - - - - - - - -

3

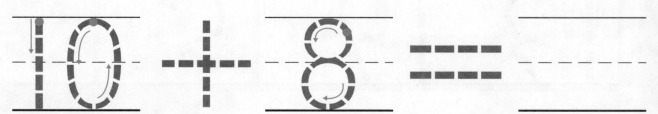

DIRECTIONS **1.** Count and tell how many. Trace the numbers. **2.** Count and tell how many. Write the number. **3.** Look at the ten frames in Exercise 2. Complete the addition sentence to match.

Name _____

19
nineteen

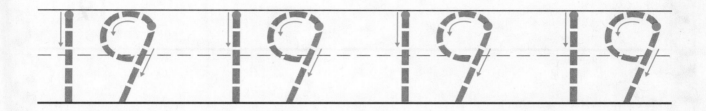

_____

- - - - - - - - - -

_____

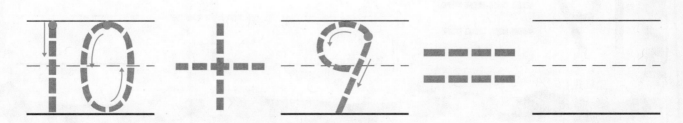

DIRECTIONS **4.** Count and tell how many. Trace the numbers. **5.** Count and tell how many. Write the number. **6.** Look at the ten frames in Exercise 5. Complete the addition sentence to match.

Chapter 7 • Lesson 10

PROBLEM SOLVING REAL WORLD

1

17

18

19

2

19 = ___ = ___ + ___

DIRECTIONS **1.** Circle a number. Draw more flowers to show that number. **2.** Draw a set of 19 objects. If you circle 10 of the objects, how many more objects are there? Complete the addition sentence to match.

HOME ACTIVITY • Ask your child to count and write the number for a set of 18 or 19 objects, such as macaroni pieces or buttons.

© Houghton Mifflin Harcourt Publishing Company

FOR MORE PRACTICE:
Standards Practice Book, pp. P145–P146

Chapter 7 Review/Test

Vocabulary

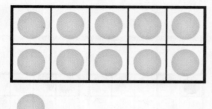 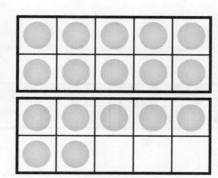 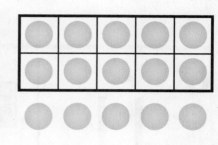

| fifteen | eleven | seventeen |

Concepts and Skills

  _____

- - - - - - - - - -

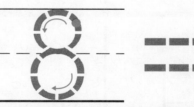 _____

- - - - - - - - - -

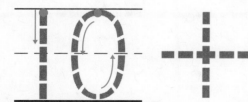

DIRECTIONS **1.** Draw lines to match the counters in the ten frames to the number word. (pp. 265, 278, 289) **2.** Count and tell how many. Write the number. **3.** Look at the ten frames in Exercise 2. Complete the addition sentence to match. **4.** Circle to show 14 beads on the bead string.

5

14

○   ○   ○   ○

6

11

○   ○   ○   ○

7

16   17   18   19

○   ○   ○   ○

8

14   15   16   17

○   ○   ○   ○

DIRECTIONS   5–6. Mark under the set that shows the number at the beginning of the row.   **7–8.** Mark under the number that shows how many.

9 🌸🌸🌸🌸🌸
🌸🌸🌸🌸🌸
🌸🌸🌸

| 13 | 15 | 17 | 18 |
| ○ | ○ | ○ | ○ |

10

| 13 | 14 | 15 | 16 |
| ○ | ○ | ○ | ○ |

11

12

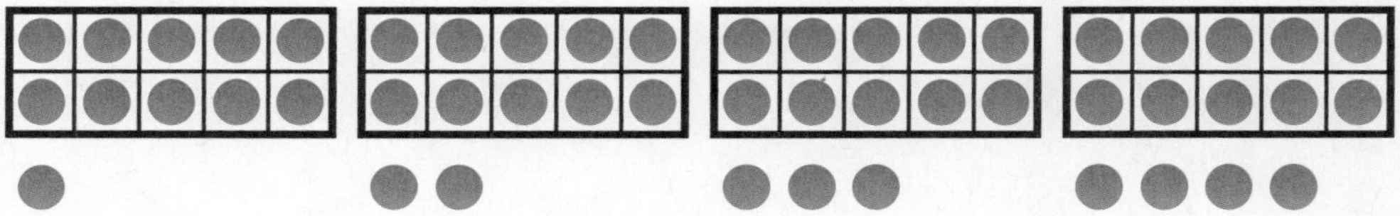

○　　　○　　　○　　　○

12

17

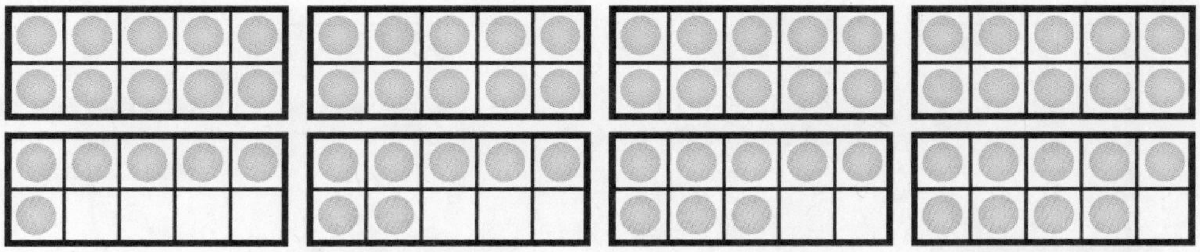

○　　　○　　　○　　　○

DIRECTIONS 9–10. Mark under the number that shows how many. **11–12.** Mark under the set that shows the number at the beginning of the row.

Performance Task

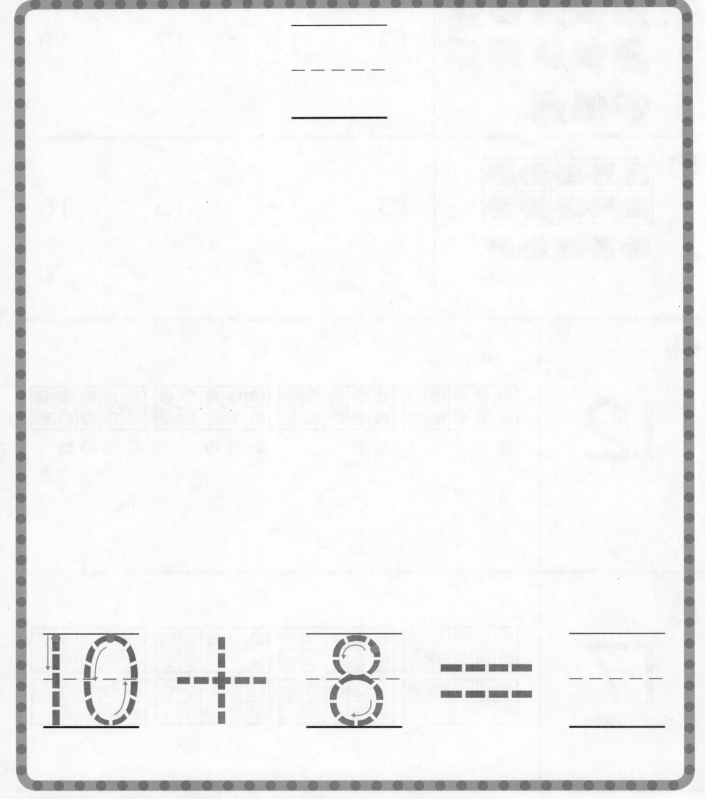

PERFORMANCE TASK This task will assess the child's understanding of the use of objects, diagrams, words, and numerals to represent the numbers 11 to 19.

Chapter 8

Represent, Count, Read, and Write 20 and Beyond

Curious About Math with Curious George

Watermelon is actually a vegetable and not a fruit.

- How many seeds can you count on this watermelon?

Name _____

Explore Numbers to 10

Compare Numbers to 10

 _____

_____

  _____

_____

Write Numbers to 10

3 _____ _____ 6 _____ 8

DIRECTIONS **1.** Circle all of the sets that show 9. **2.** Circle all of the sets that show 8. **3.** Count and tell how many. Write the number. Circle the number that is less. **4.** Write the numbers in order as you count forward.

FAMILY NOTE: This page checks your child's understanding of important skills needed for success in Chapter 8.

GO Online Assessment Options Soar to Success Math

Vocabulary Builder

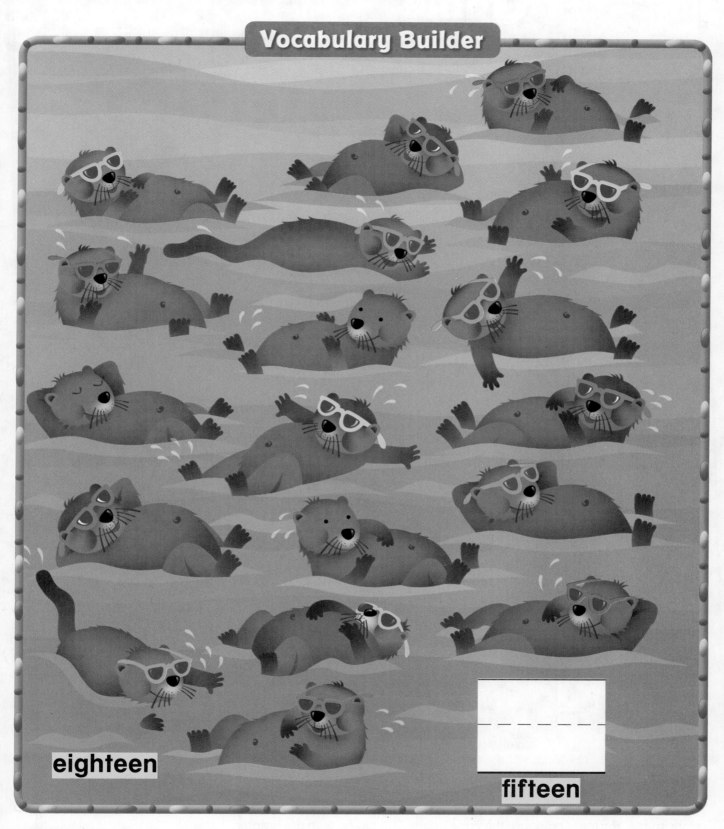

eighteen

fifteen

DIRECTIONS Point to each otter as you count. Point to the number word that shows how many otters in all. How many are wearing glasses? Write the number.

• eStudent Edition
• Multimedia eGlossary

Chapter 8

Who Has More?

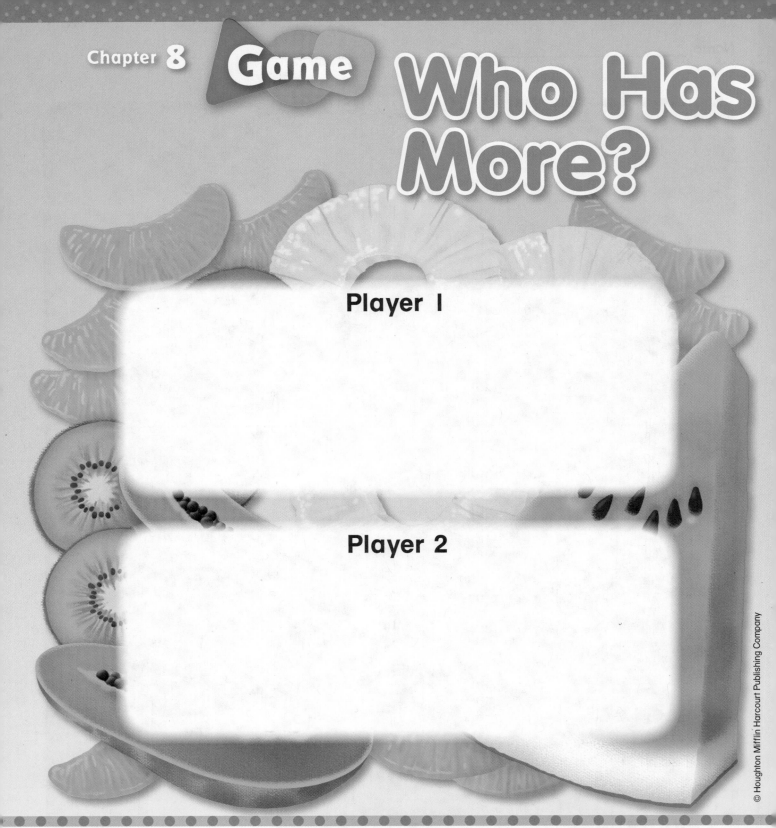

Player 1

Player 2

DIRECTIONS Play with a partner. Each player shuffles a set of numeral cards and places them facedown in a stack. Each player turns over the top card on his or her stack and models that number by placing cube trains on the work space. Partners compare the cube trains. The player with the greater number keeps both of the numeral cards. If both numbers are the same, each player returns the card to the bottom of his or her stack. The player with the most cards at the end of the game wins.

MATERIALS 2 sets of numeral cards 11–20, cubes

Name _____

Model and Count 20

Essential Question How can you show and count 20 objects?

Listen and Draw

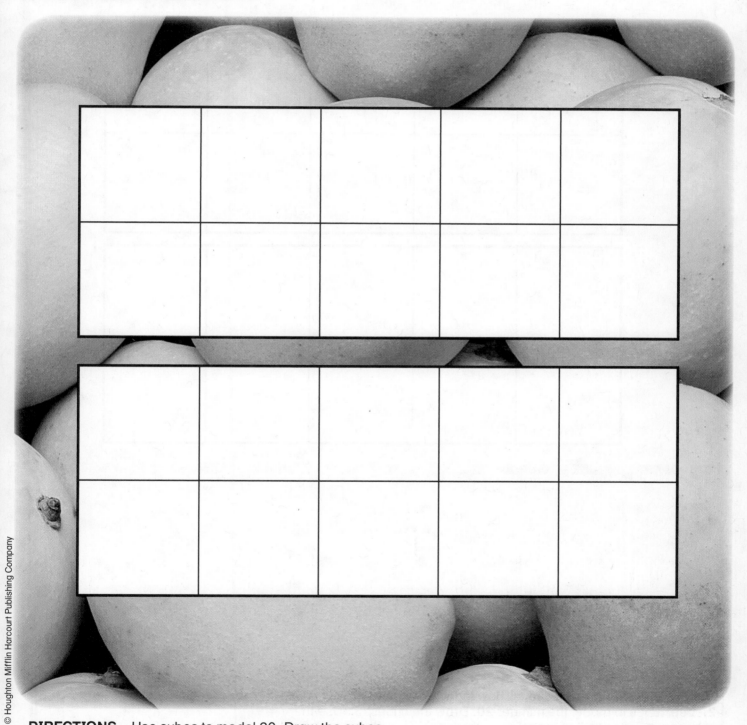

DIRECTIONS Use cubes to model 20. Draw the cubes.

Share and Show

 20
twenty

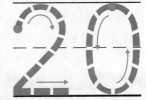

 2

3 ✓

DIRECTIONS 1. Count and tell how many. Trace the number.
2. Use cubes to model the number 20. Draw the cubes. 3. Use the
cubes from Exercise 2 to model ten-cube trains. Draw the cube trains.

310 three hundred ten

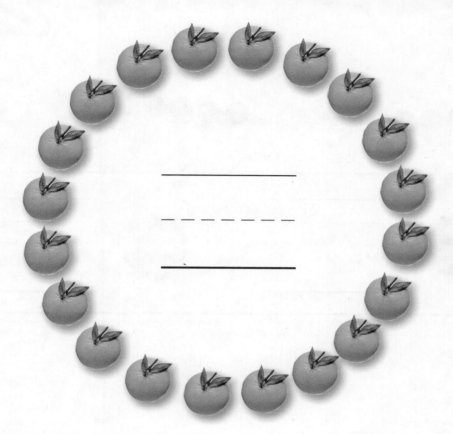

_____

- - - - - - - - -

_____

_____

- - - - - - - - -

_____

DIRECTIONS 4–5. Count and tell how many pieces of fruit. Write the number. Tell a friend how you counted the oranges.

PROBLEM SOLVING REAL WORLD

①

②

③

DIRECTIONS 1. Circle to show 20 beads. **2.** How many of each color bead did you circle? Write the numbers. Tell a friend about the number of each color beads. **3.** Draw and write to show what you know about 20. Tell a friend about your drawing.

HOME ACTIVITY • Draw two ten frames on a sheet of paper. Have your child show the number 20 by placing small objects, such as buttons or dried beans, in the ten frames.

FOR MORE PRACTICE:
Standards Practice Book, pp. P151–P152

Read and Write to 20

Essential Question How can you read and write 20 with words and numbers?

Listen and Draw

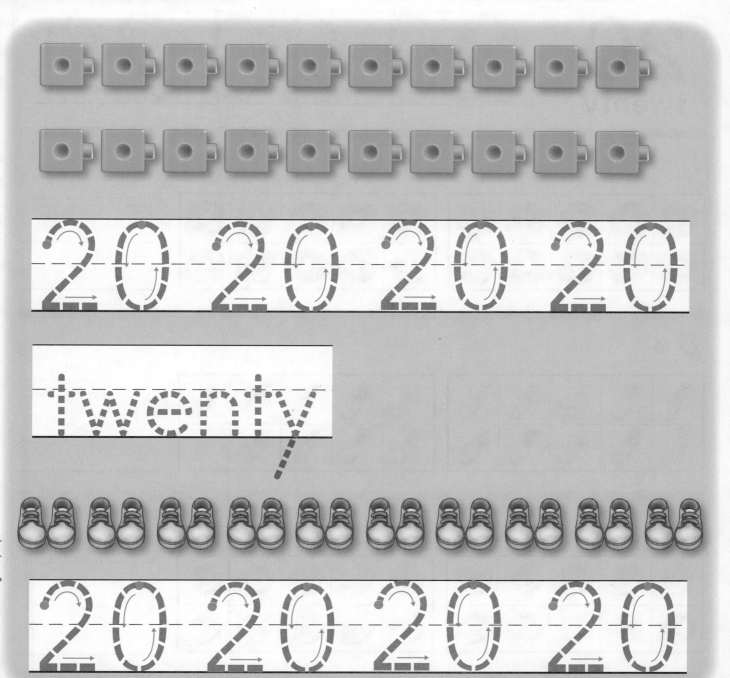

DIRECTIONS Count and tell how many cubes. Trace the numbers and the word. Count and tell how many shoes. Trace the numbers.

Chapter 8 • Lesson 2

three hundred thirteen **313**

Share and Show

1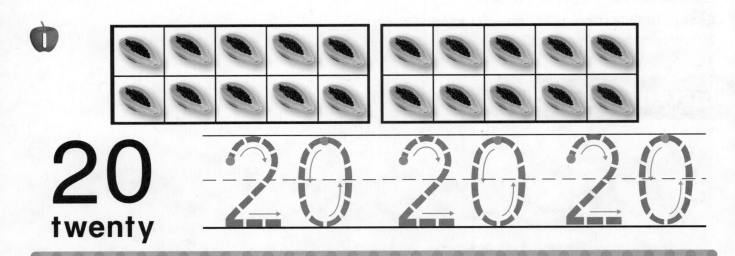

20
twenty

2

_____
- - - - - - - - - -
_____

3

_____
- - - - - - - - - -
_____

4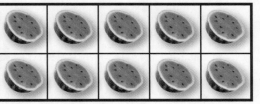

_____
- - - - - - - - - -
_____

DIRECTIONS **1.** Count and tell how many pieces of fruit. Trace the numbers as you say them. **2–4.** Count and tell how many pieces of fruit. Write the number.

Name _____

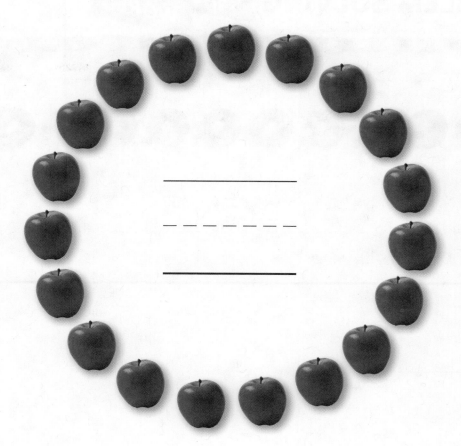

_____

_ _ _ _ _ _ _

_____

_____

_ _ _ _ _ _ _

_____

DIRECTIONS **5–6.** Count and tell how many pieces of fruit. Write the number.

© Houghton Mifflin Harcourt Publishing Company

PROBLEM SOLVING REAL WORLD

1

18

19

20

2

DIRECTIONS **I.** Circle a number. Draw more fruit to show that number. **2.** Draw a set of objects that has a number of objects one greater than 19. Write how many objects are in the set. Tell a friend about your drawing.

HOME ACTIVITY • Have your child use small objects, such as pebbles or pasta pieces, to show the number 20. Then have him or her write the number on a piece of paper.

FOR MORE PRACTICE:
Standards Practice Book, pp. P153–P154

Name _____

Count and Order to 20

Essential Question How can you count forward to 20 from a given number?

Listen and Draw

1 2 3 4 5 6 7 8 9 10 11 12 13 14 15 16 17 18 19 20

DIRECTIONS Draw a line under a number. Count forward to 20 from that number. Use the terms *greater than* and *less than* to compare and describe the order of numbers. Circle the number that is one greater than the number you underlined. Build cube trains to model the numbers you marked. Draw the cube trains. Circle the larger cube train.

Chapter 8 • Lesson 3

Share and Show

1 2 3 4 5 6 7 8 9 10 11 12 13 14 15 16 17 18 19 20

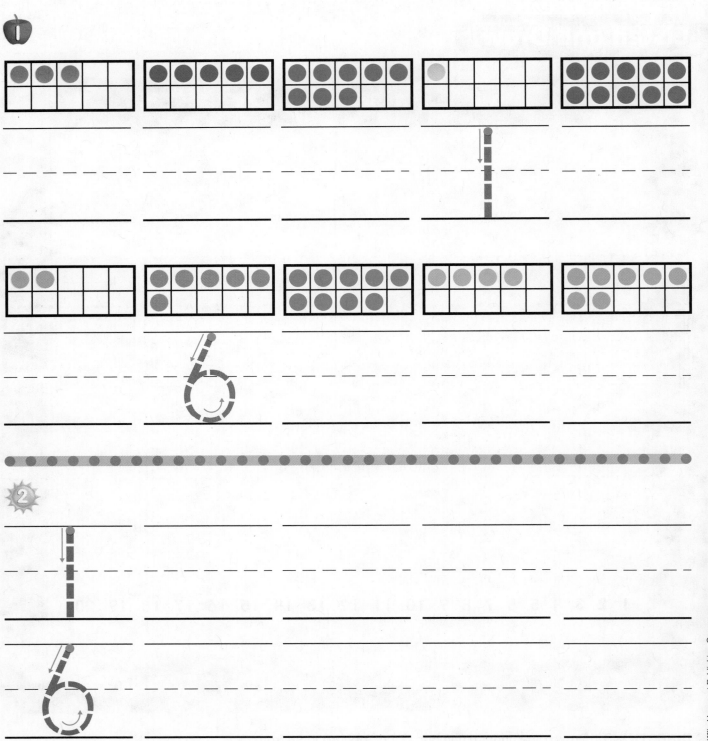

DIRECTIONS **1.** Count the dots of each color in the ten frames. Trace or write the numbers. **2.** Trace and write those numbers in order.

3 ☑

(ten frames row 1 and 2)

‾ ‾

16

(ten frames rows)

‾ ‾

4

16

DIRECTIONS **3.** Count the dots of each color in the ten frames.
Trace or write the numbers. **4.** Trace and write those numbers in order.

PROBLEM SOLVING REAL WORLD

1	2	----	4	5
6	7	8	9	----
11	----	13	14	15
16	17	----	19	20

DIRECTIONS Write to show the numbers in order. Count forward to 20 from one of the numbers you wrote.

HOME ACTIVITY • Give your child a set of 11 objects, a set of 12 objects, and a set of 13 objects. Have him or her count the objects in each set and place the sets in numerical order.

FOR MORE PRACTICE:
Standards Practice Book, pp. P155–P156

© Houghton Mifflin Harcourt Publishing Company

Name _____

Problem Solving • Compare Numbers to 20

Essential Question How can you solve problems using the strategy *make a model*?

🔑 Unlock the Problem

DIRECTIONS Alison has a number of yellow cubes one greater than 15. Josh has a number of green cubes one less than 17. Show the cubes. Compare the sets of cubes. Draw the cubes. Tell a friend about your drawing.

Chapter 8 • Lesson 4

three hundred twenty-one **321**

Try Another Problem

1 ✓

_____

- - - - - - - - - -

_____

_____

- - - - - - - - - -

_____

DIRECTIONS 1. Kaelin has 18 apples. She has a number of apples two greater than Chase. Use cubes to model the sets of apples. Compare the sets. Which set is larger? Draw the cubes. Write how many in each set. Circle the greater number. Tell a friend how you compared the numbers.

Name _____

Share and Show

_____

- - - - - - -

_____

_____

- - - - - - -

_____

DIRECTIONS 2. Skyler has 19 oranges. Taylor has a number of oranges two less than Skyler. Use cubes to model the sets of oranges. Compare the sets. Which set is smaller? Draw the cubes. Write how many in each set. Circle the number that is less. Tell a friend how you compared the numbers.

HOME ACTIVITY • Have your child count two sets of objects in your home, and write how many are in each set. Then have him or her circle the greater number. Repeat with sets of different numbers.

FOR MORE PRACTICE:
Standards Practice Book, pp. P157–P158

© Houghton Mifflin Harcourt Publishing Company

Concepts and Skills

_____

_____

_____

_____

_____

_____

_____

_____

_____

17	18	19	20
○	○	○	○

DIRECTIONS   1. Count and tell how many. Write the number.
2. Write how many pieces of fruit are in each picture. Circle the number that is less.   **3.** Write how many pieces of fruit are in each picture. Circle the number that is greater.   **4.** Mark under the number that shows how many pieces of fruit are at the beginning of the row.

Count to 50 by Ones

Essential Question How does the order of
numbers help you count to 50 by ones?

Listen and Draw

1	2	3	4	5	6	7	8	9	10
11	12	13	14	15	16	17	18	19	20
21	22	23	24	25	26	27	28	29	30
31	32	33	34	35	36	37	38	39	40
41	42	43	44	45	46	47	48	49	50

DIRECTIONS Point to each number as you count to 50. Trace the
circle around the number 50.

Chapter 8 · Lesson 5

three hundred twenty-five **325**

Share and Show

1	2	3	4	5	6	7	8	9	10
11	12	13	14	15	16	17	18	19	20
21	22	23	24	25	26	27	28	29	30
31	32	33	34	35	36	37	38	39	40
41	42	43	44	45	46	47	48	49	50

DIRECTIONS **I.** Point to each number as you count to 50. Circle the number 15. Begin with 15 and count forward to 50. Draw a line under the number 50.

1	2	3	4	5	6	7	8	9	10
11	12	13	14	15	16	17	18	19	20
21	22	23	24	25	26	27	28	29	30
31	32	33	34	35	36	37	38	39	40
41	42	43	44	45	46	47	48	49	50

DIRECTIONS **2.** Look away and point to any number. Circle that number. Count forward from that number. Draw a line under the number 50.

Chapter 8 • Lesson 5   three hundred twenty-seven **327**

PROBLEM SOLVING

1	2	3	4	5	6	7	8	9	10
11	12	13	14	15	16	17	18	19	20
21	22	23	24	25	26	27	28	29	30
31	32	33	34	35	36	37	38	39	40
41	42	43	44	45	46	47	48	49	50

DIRECTIONS I am greater than 17 and less than 19. What number am I? Use blue to color that number. I am greater than 24 and less than 26. What number am I? Use red to color that number.

HOME ACTIVITY • Think of a number between 1 and 50. Say *greater than* and *less than* to describe your number. Have your child say the number.

FOR MORE PRACTICE:
Standards Practice Book, pp. P159–P160

Name _____

Count to 100 by Ones

Essential Question How does the order of numbers help you count to 100 by ones?

Listen and Draw

1	2	3	4	5	6	7	8	9	10
11	12	13	14	15	16	17	18	19	20
21	22	23	24	25	26	27	28	29	30
31	32	33	34	35	36	37	38	39	40
41	42	43	44	45	46	47	48	49	50
51	52	53	54	55	56	57	58	59	60
61	62	63	64	65	66	67	68	69	70
71	72	73	74	75	76	77	78	79	80
81	82	83	84	85	86	87	88	89	90
91	92	93	94	95	96	97	98	99	100

DIRECTIONS Point to each number as you count to 100. Trace the circle around the number 100.

Share and Show

1	2	3	4	5	6	7	8	9	10
11	12	13	14	15	16	17	18	19	20
21	22	23	24	25	26	27	28	29	30
31	32	33	34	35	36	37	38	39	40
41	42	43	44	45	46	47	48	49	50
51	52	53	54	55	56	57	58	59	60
61	62	63	64	65	66	67	68	69	70
71	72	73	74	75	76	77	78	79	80
81	82	83	84	85	86	87	88	89	90
91	92	93	94	95	96	97	98	99	100

DIRECTIONS **1.** Point to each number as you count to 100.
Circle the number 11. Begin with 11 and count forward to 100.
Draw a line under the number 100.

1	2	3	4	5	6	7	8	9	10
11	12	13	14	15	16	17	18	19	20
21	22	23	24	25	26	27	28	29	30
31	32	33	34	35	36	37	38	39	40
41	42	43	44	45	46	47	48	49	50
51	52	53	54	55	56	57	58	59	60
61	62	63	64	65	66	67	68	69	70
71	72	73	74	75	76	77	78	79	80
81	82	83	84	85	86	87	88	89	90
91	92	93	94	95	96	97	98	99	100

DIRECTIONS **2.** Point to each number as you count to 100. Look away and point to any number. Circle that number. Count forward to 100 from that number. Draw a line under the number 100.

PROBLEM SOLVING

1	2	3	4	_____	6	7	8	9	10
11	12	13	_____	15	_____	17	18	19	20
21	22	23	24	_____	26	27	28	29	30

②

DIRECTIONS 1. Place your finger on the number 15. Write to show the numbers that are "neighbors" to the number 15. Say *greater than* and *less than* to describe the numbers. 2. Draw to show what you know about some other "neighbor" numbers in the chart.

HOME ACTIVITY • Show your child a calendar. Point to a number on the calendar. Have him or her tell you all the numbers that are "neighbors" to that number.

FOR MORE PRACTICE:
Standards Practice Book, pp. P161–P162

Name _____

Count to 100 by Tens

Essential Question How can you count to 100 by
tens on a hundred chart?

Listen and Draw

1	2	3	4	5	6	7	8	9	10
11	12	13	14	15	16	17	18	19	20
21	22	23	24	25	26	27	28	29	30
31	32	33	34	35	36	37	38	39	40
41	42	43	44	45	46	47	48	49	50
51	52	53	54	55	56	57	58	59	60
61	62	63	64	65	66	67	68	69	70
71	72	73	74	75	76	77	78	79	80
81	82	83	84	85	86	87	88	89	90
91	92	93	94	95	96	97	98	99	100

DIRECTIONS Trace the circles around the numbers that end in a 0.
Beginning with 10, count those numbers in order. Tell a friend how you
are counting.

Chapter 8 • Lesson 7   three hundred thirty-three **333**

Share and Show

1	2	3	4	5	6	7	8	9	_____
11	12	13	14	15	16	17	18	19	_____
21	22	23	24	25	26	27	28	29	30
31	32	33	34	35	36	37	38	39	40
41	42	43	44	45	46	47	48	49	50

DIRECTIONS **1.** Write the numbers to complete the counting order to 20. Trace the numbers to complete the counting order to 50. Count by tens as you point to the numbers you wrote and traced.

© Houghton Mifflin Harcourt Publishing Company

51	52	53	54	55	56	57	58	59	60
61	62	63	64	65	66	67	68	69	70
71	72	73	74	75	76	77	78	79	80
81	82	83	84	85	86	87	88	89	90
91	92	93	94	95	96	97	98	99	100

DIRECTIONS 2. Trace the numbers to complete the counting order to 100. Count by tens as you point to the numbers you traced.

PROBLEM SOLVING

1	2	3	4	5	6	7	8	9	----
11	12	13	14	15	16	17	18	19	----
21	22	23	24	25	26	27	28	29	30
31	32	33	34	35	36	37	38	39	40
41	42	43	44	45	46	47	48	49	50

DIRECTIONS Tony has 10 marbles. Write the number in order. Jenny has ten more marbles than Tony. Write that number in order. Lindsay has ten more marbles than Jenny. Draw a line under the number that shows how many marbles Lindsay has. When counting by tens, what number comes right after 40? Circle the number.

HOME ACTIVITY • Show your child a calendar. Use self-stick notes to cover random numbers. Ask your child to say the numbers that are covered. Then have him or her remove the self-stick note to check.

FOR MORE PRACTICE:
Standards Practice Book, pp. P163–P164

Name _____

Count by Tens

Essential Question How can you use sets of tens to count to 100?

Listen and Draw REAL WORLD

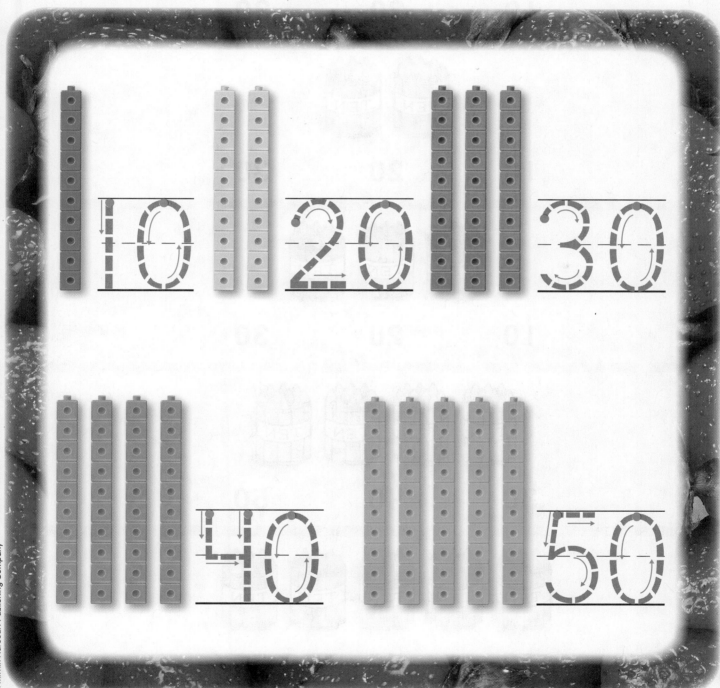

DIRECTIONS Point to each set of cube towers as you count by tens. Trace the numbers as you count by tens.

Chapter 8 • Lesson 8

three hundred thirty-seven **337**

Share and Show

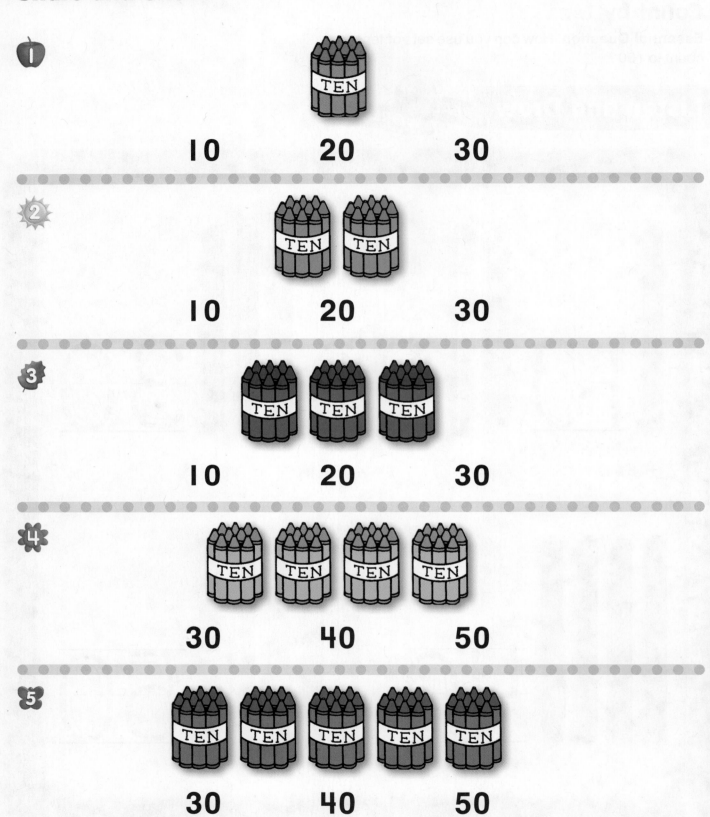

1

10   **20**   30

2

10   **20**   30

3

10   **20**   30

4

30   **40**   50

5

30   **40**   50

DIRECTIONS 1–5. Point to each set of 10 as you count by tens. Circle the number that shows how many.

Name _____

 6 ✓

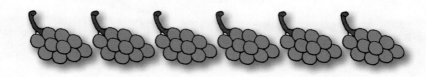

60   **70**   **80**

7 ✓

60   **70**   **80**

8

80   **90**   **100**

9

80   **90**   **100**

10

80   **90**   **100**

DIRECTIONS 6–10. Point to each set of 10 as you count by tens.
Circle the number that shows how many.

PROBLEM SOLVING REAL WORLD

DIRECTIONS Circle sets of 10 stars.
Count the sets of stars by tens.

HOME ACTIVITY • Give your child some
macaroni or dried beans and ten cups. Ask him
or her to place ten pieces of macaroni into each
cup. Then have him or her point to each cup as
he or she counts by tens to 100.

340 three hundred forty

FOR MORE PRACTICE:
Standards Practice Book, pp. P165–P166

Chapter 8 Review/Test

Vocabulary

 1

twenty

Concepts and Skills

2

_____

- - - - - - - - - -

_____

3

_____

15 - - - - - - - - - - 20

_____

4

1	2	3	4	5	6	7	8	9	10
11	12	13	14	15	16	17	18	19	20
21	22	23	24	25	26	27	28	29	30

DIRECTIONS 1. Circle beads to match the number word. (p. 312) 2. Count and tell how many pieces of fruit. Write the number. 3. Start with 15. Count forward. Write the numbers in order. 4. Color each number on the chart that ends in a 0. Count those numbers by tens in order.

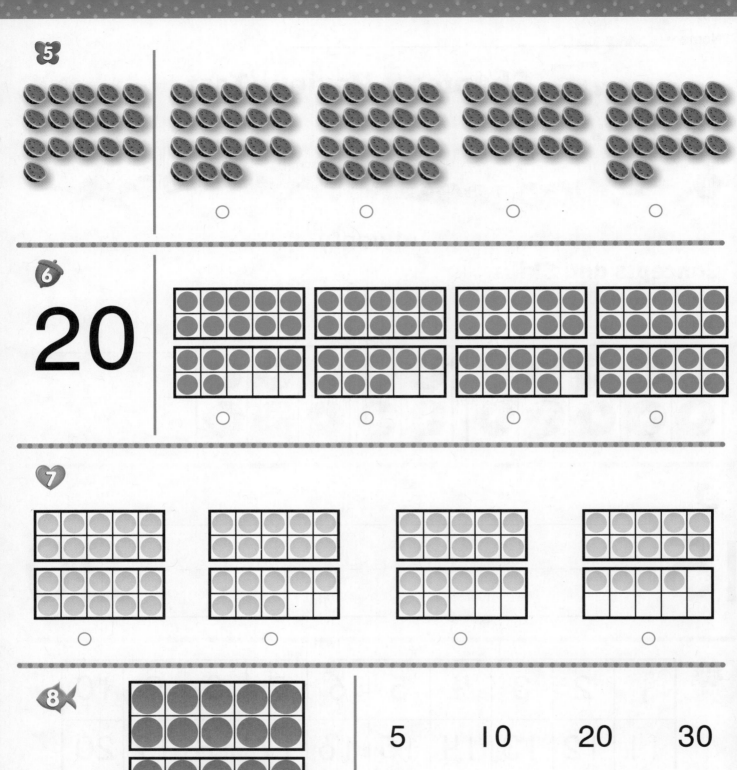

DIRECTIONS **5.** Mark under the set that has a number of watermelons less than the set at the beginning of the row. **6.** Mark under the set that models the number at the beginning of the row. **7.** Mark under the set that has a number of counters greater than 19. **8.** Mark under the number that shows how many.

9 11 12 10   13 17 15   14 15 16   15 14 19
       ○   ○   ○   ○

10

1	2	3	4	5	6	7	8	9	10
11	12	13	14	15	16	17	18	19	20
21	22	23	24	25	26	27	28	29	30
31	32	33	34	35	36	37	38	39	40
41	42	43	44	45	46	47	48	49	

   40   50   60   70
   ○   ○   ○   ○

11

        ○   ○   ○   ○

12

   50   60   70   80
   ○   ○   ○   ○

DIRECTIONS **9.** Mark under the numbers that show them in order.
10. Point to each number as you count. Mark under the number that
completes the counting order.   **11.** Mark under the set that has a number
of apples greater than the set at the beginning of the row.   **12.** Count the
crayons by tens. Mark under the number that shows how many.

Performance Task

_____

- - - - - - -

_____

_____

- - - - - - -

_____

_____

- - - - - - -

_____

PERFORMANCE TASK This task will assess the child's understanding of identifying and ordering numbers to 20.

School Fun

written by Ann Dickson

**Describing shapes
and space**

1. Sign in.

2. Put your book bag away.

3. Choose a center.

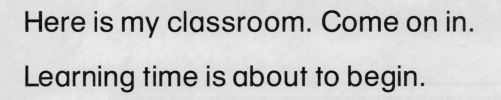

Here is my classroom. Come on in.

Learning time is about to begin.

Social Studies

Why do we have rules?

These are the book bags

we hang by our names.

Circle the ones that look the same.

Social Studies

Why do we need to take turns?

Here are the books. We read them all!

Which books are big?

Which books are small?

Social Studies

Why do we help others?

Here are markers of every kind.

Name all of the colors that you can find.

Social Studies

Why do we put things away?

Our blocks and toys are over there.

Which shapes are round?

Which shapes are square?

Social Studies

Why do we share?

Write About the Story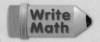

> **Vocabulary Review**
>
> alike
>
> different

DIRECTIONS These lunch boxes are alike. In one lunch box draw something that you like to eat. Now circle the lunch box that is different.

Alike and Different

DIRECTIONS **1.** Color the markers so that they match the colors of the cups.
2. Color the book bags that are alike by shape. **3.** This classroom needs some
books. Draw a book that is a different size.

Identify and Describe Two-Dimensional Shapes

Curious About Math with Curious George

The sails on these boats are shaped like a triangle.

- How many stripes can you count on the first sail?

Name _____

Show What You Know

Shape

 1

 2

 3

Count Objects

 4

_____

- - - - - - - -

_____

 5

_____

- - - - - - - -

_____

DIRECTIONS **1–3.** Look at the shape at the beginning of the row. Mark an X on the shape that is alike. **4–6.** Count and tell how many. Write the number.

FAMILY NOTE: This page checks your child's understanding of important skills needed for success in Chapter 9.

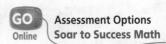 **GO** Online Assessment Options Soar to Success Math

© Houghton Mifflin Harcourt Publishing Company

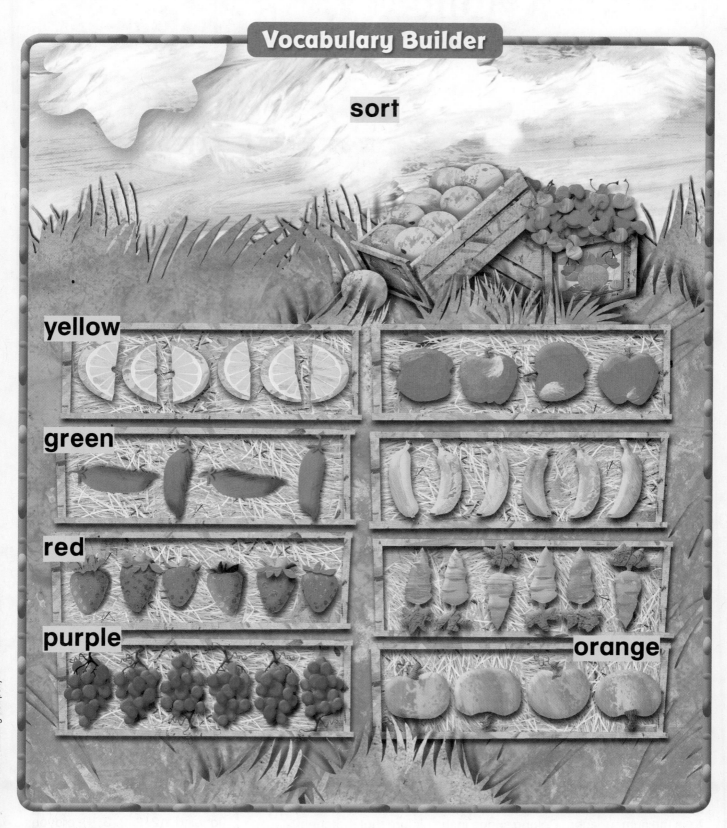

Vocabulary Builder

sort

yellow

green

red

purple

orange

DIRECTIONS Circle the box that is sorted by green vegetables. Mark an X on the box that is sorted by purple fruit.

• eStudent Edition
• Multimedia *eGlossary*

Game

Number Picture

DIRECTIONS Play with a partner. Decide who goes first. Toss the number cube. Color a shape in the picture that matches the number rolled. A player misses a turn if a number is rolled and all shapes with that number are colored. Continue until all shapes in the picture are colored.

MATERIALS number cube (labeled 1,2, 2, 3, 3, 4), crayons

© Houghton Mifflin Harcourt Publishing Company

Identify and Name Circles

Essential Question How can you identify and name circles?

Listen and Draw  REAL WORLD

circles	not circles

DIRECTIONS Place two-dimensional shapes on the page. Identify and name the circles. Sort the shapes by circles and not circles. Trace and color the shapes on the sorting mat.

Share and Show

DIRECTIONS 1. Mark an X on all of the circles.

Name _____

DIRECTIONS 2. Color the circles in the picture.

Chapter 9 • Lesson 1

PROBLEM SOLVING

2

DIRECTIONS 1. Which shape is a circle? Mark an X on that shape. 2. Draw to show what you know about circles. Tell a friend about your drawing.

HOME ACTIVITY · Have your child show you an object that is shaped like a circle.

FOR MORE PRACTICE:
Standards Practice Book, pp. P171–P172

Describe Circles

Essential Question How can you describe circles?

© Houghton Mifflin Harcourt Publishing Company

Listen and Draw REAL WORLD

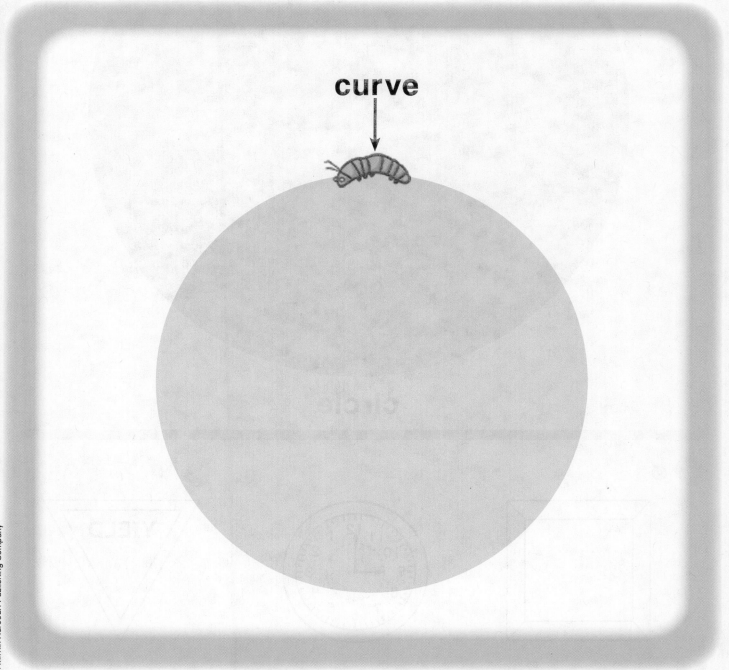

curve

DIRECTIONS Use your finger to trace around the circle.
Talk about the curve. Trace around the curve.

circle

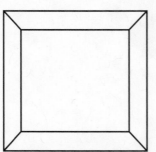

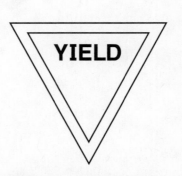

DIRECTIONS 1. Use your finger to trace around the circle. Trace the curve around the circle. 2. Color the object that is shaped like a circle.

Name _____

DIRECTIONS **3.** Use a pencil to hold one end of a large paper clip on one of the dots in the center of the page. Place another pencil in the other end of the paper clip. Move the pencil around to draw a circle.

Chapter 9 • Lesson 2

PROBLEM SOLVING

DIRECTIONS I have a curve. What shape am I? Draw the shape. Tell a friend the name of the shape.

HOME ACTIVITY • Have your child describe a circle.

364 three hundred sixty-four

Identify and Name Squares

Essential Question How can you identify and name squares?

Listen and Draw REAL WORLD

squares	not squares

DIRECTIONS Place two-dimensional shapes on the page. Identify and name the squares. Sort the shapes by squares and not squares. Trace and color the shapes on the sorting mat.

Share and Show

DIRECTIONS I. Mark an X on all of the squares.

Name _____

DIRECTIONS 2. Color the squares in the picture.

Chapter 9 • Lesson 3

three hundred sixty-seven **367**

PROBLEM SOLVING

1

2

DIRECTIONS 1. Which shapes are squares? Mark an X on those shapes. 2. Draw to show what you know about squares. Tell a friend about your drawing.

HOME ACTIVITY • Have your child show you an object that is shaped like a square.

368 three hundred sixty-eight

FOR MORE PRACTICE: Standards Practice Book, pp. P175–P176

Name _____

Describe Squares

Essential Question How can you describe squares?

Listen and Draw

vertex

side

DIRECTIONS Use your finger to trace around the square. Talk about the number of sides and the number of vertices. Draw an arrow pointing to another vertex. Trace around the sides.

Chapter 9 • Lesson 4

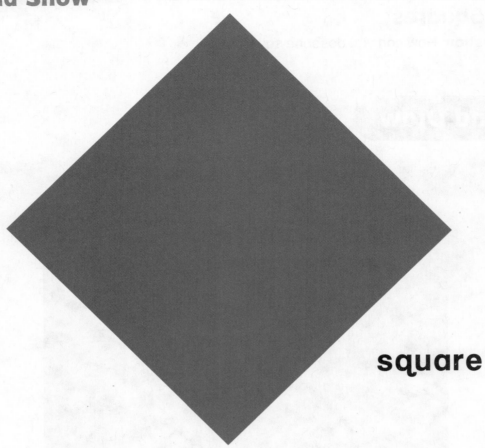

square

 ☑ ___ ___

vertices

② ☑ ___ ___

sides

DIRECTIONS 1. Place a counter on each corner, or vertex. Write how many corners, or vertices. 2. Trace around the sides. Write how many sides.

Name _____

DIRECTIONS 3. Draw and color a square.

PROBLEM SOLVING

DIRECTIONS I have 4 sides of equal length and 4 vertices. What shape am I? Draw the shape. Tell a friend the name of the shape.

HOME ACTIVITY • Have your child describe a square.

FOR MORE PRACTICE:
Standards Practice Book, pp. P177–P178

Identify and Name Triangles

Essential Question How can you identify
and name triangles?

Listen and Draw REAL WORLD

triangles	not triangles

DIRECTIONS Place two-dimensional shapes on the page.
Identify and name the triangles. Sort the shapes by triangles and not
triangles. Trace and color the shapes on the sorting mat.

Share and Show

Name _____

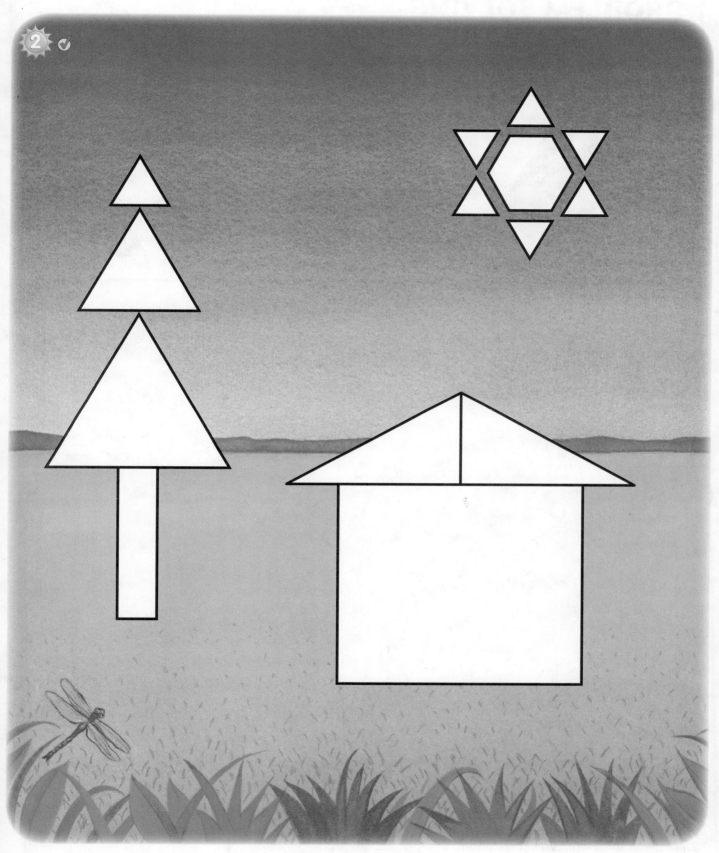

DIRECTIONS 2. Color the triangles in the picture.

PROBLEM SOLVING

DIRECTIONS   **I.** Which shapes are triangles? Mark an X on those shapes. **2.** Draw to show what you know about triangles. Tell a friend about your drawing.

HOME ACTIVITY • Have your child show you an object that is shaped like a triangle.

FOR MORE PRACTICE:
Standards Practice Book, pp. P179–P180

Name _____

Describe Triangles

Essential Question How can you describe triangles?

Listen and Draw

vertex

side

DIRECTIONS Use your finger to trace around the triangle. Talk about the number of sides and the number of vertices. Draw an arrow pointing to another vertex. Trace around the sides.

Chapter 9 • Lesson 6

three hundred seventy-seven **377**

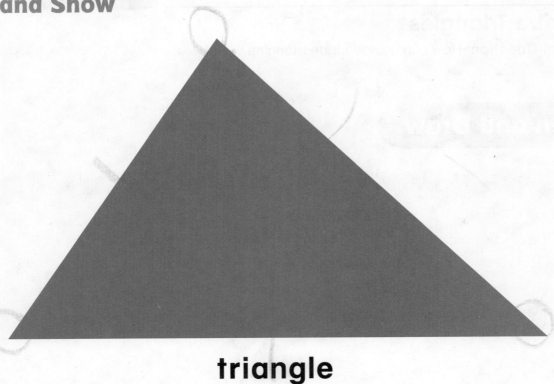

triangle

 _____

_____

vertices

_____

_____

sides

DIRECTIONS **I.** Place a counter on each corner, or vertex. Write how many corners, or vertices. **2.** Trace around the sides. Write how many sides.

Name _____

DIRECTIONS **3.** Draw and color a triangle.

HOME ACTIVITY • Have your child describe a triangle.
FOR MORE PRACTICE:
Standards Practice Book, pp. P181–P182

Chapter 9 • Lesson 6

✓ Mid-Chapter Checkpoint

Concepts and Skills

1 ___4___ sides

_____ vertices

2 _____ sides

_____ vertices

3

4

5

○   ○   ○   ○

© Houghton Mifflin Harcourt Publishing Company

DIRECTIONS 1–2. Trace around each side. Write how many sides. Place a counter on each corner or vertex. Write how many vertices. **3.** Mark an X on all of the circles. **4.** Mark an X on all of the triangles. **5.** Mark under the shape that is a triangle.

Identify and Name Rectangles

Essential Question How can you identify and name rectangles?

Listen and Draw  REAL WORLD

rectangles	not rectangles

DIRECTIONS Place two-dimensional shapes on the page. Identify and name the rectangles. Sort the shapes by rectangles and not rectangles. Trace and color the shapes on the sorting mat.

Share and Show

DIRECTIONS **1.** Mark an X on all of the rectangles.

Name _____

DIRECTIONS 2. Color the rectangles in the picture.

PROBLEM SOLVING

1

2

DIRECTIONS **1.** Which shapes are rectangles? Mark an X on those shapes. **2.** Draw to show what you know about rectangles. Tell a friend about your drawing.

HOME ACTIVITY • Have your child show you an object that is shaped like a rectangle.

FOR MORE PRACTICE:
Standards Practice Book, pp. P183–P184

Name _____

Describe Rectangles

Essential Question How can you describe rectangles?

Listen and Draw

side

vertex

DIRECTIONS Use your finger to trace around the rectangle. Talk about the number of sides and the number of vertices. Draw an arrow pointing to another vertex. Trace around the sides.

Chapter 9 · Lesson 8

three hundred eighty-five **385**

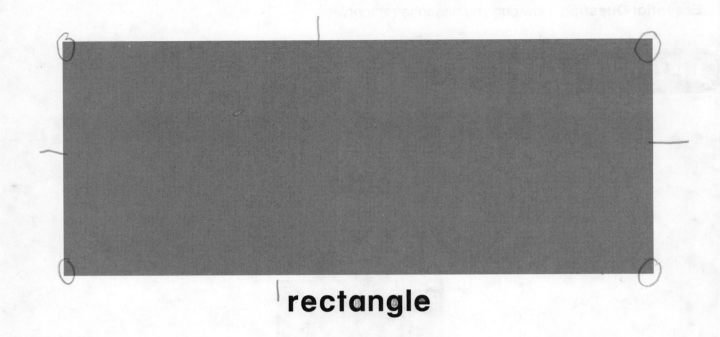

rectangle

  **vertices**

 sides

DIRECTIONS 1. Place a counter on each corner, or vertex. Write how many corners, or vertices. 2. Trace around the sides. Write how many sides.

DIRECTIONS 3. Draw and color a rectangle.

PROBLEM SOLVING

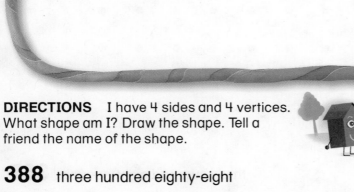

DIRECTIONS   I have 4 sides and 4 vertices. What shape am I? Draw the shape. Tell a friend the name of the shape.

HOME ACTIVITY • Have your child describe a rectangle.

388   three hundred eighty-eight

FOR MORE PRACTICE:
Standards Practice Book, pp. P185–P186

Identify and Name Hexagons

Essential Question How can you identify and name hexagons?

Listen and Draw

hexagons	not hexagons

DIRECTIONS Place two-dimensional shapes on the page. Identify and name the hexagons. Sort the shapes by hexagons and not hexagons. Trace and color the shapes on the sorting mat.

Share and Show

1

Name _____

DIRECTIONS 2. Color the hexagons in the picture.

PROBLEM SOLVING

① ②

DIRECTIONS 1. Which shapes are hexagons? Mark an X on those shapes. 2. Draw to show what you know about hexagons. Tell a friend about your drawing.

HOME ACTIVITY • Draw some shapes on a page. Include several hexagons. Have your child circle the hexagons.

FOR MORE PRACTICE:
Standards Practice Book, pp. P187–P188

Describe Hexagons

Essential Question How can you describe hexagons?

Listen and Draw

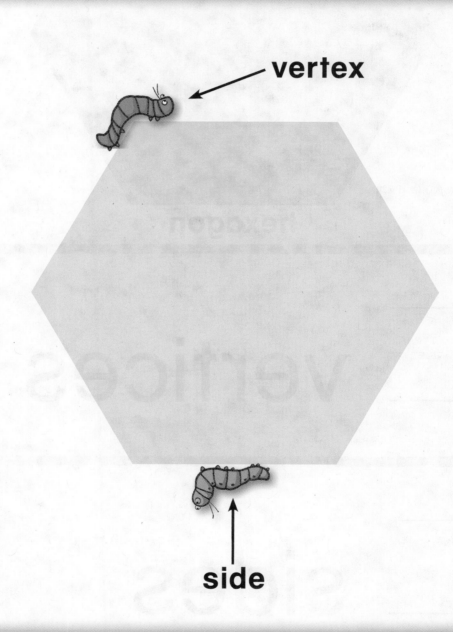

vertex

side

DIRECTIONS Use your finger to trace around the hexagon. Talk about the number of sides and the number of vertices. Draw an arrow pointing to another vertex. Trace around the sides.

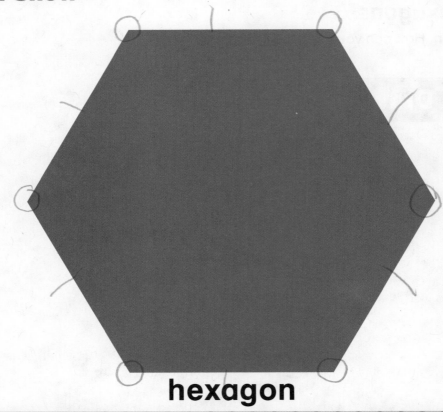

hexagon

 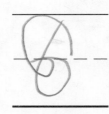 vertices

  sides

DIRECTIONS   1. Place a counter on each corner, or vertex. Write how many corners, or vertices.   **2.** Trace around the sides. Write how many sides.

Name _____

DIRECTIONS **3.** Draw and color a hexagon.

PROBLEM SOLVING

DIRECTIONS I have 6 sides and 6 vertices. What shape am I? Draw the shape. Tell a friend the name of the shape.

HOME ACTIVITY • Have your child describe a hexagon.

396 three hundred ninety-six

FOR MORE PRACTICE: Standards Practice Book, pp. P189–P190

Name _____

Algebra • Compare Two-Dimensional Shapes

Essential Question How can you use the words *alike* and *different* to compare two-dimensional shapes?

Listen and Draw

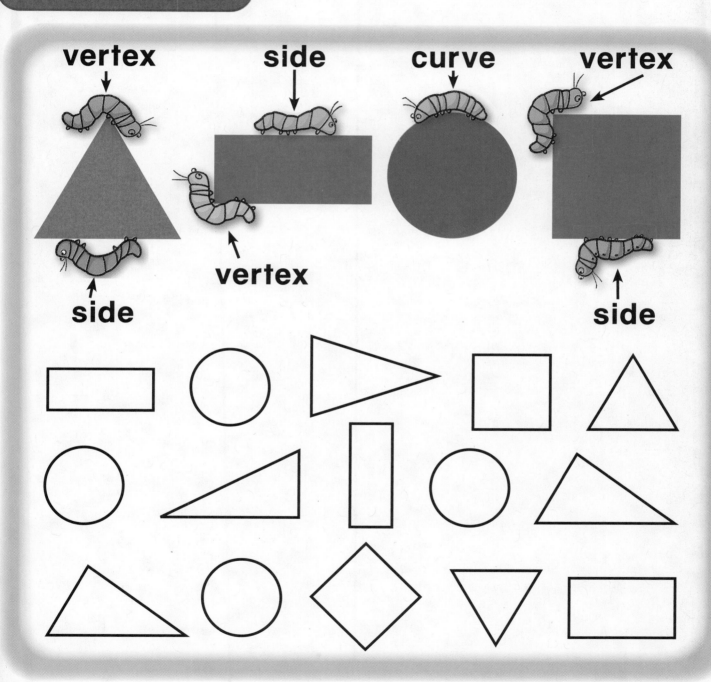

vertex   side   curve   vertex

side   vertex   side

DIRECTIONS Look at the worms and the shapes. Use the words *alike* and *different* to compare the shapes. Use green to color the shapes with four vertices and four sides. Use blue to color the shapes with curves. Use red to color the shapes with three vertices and three sides.

Share and Show

alike	different

DIRECTIONS **1.** Place two-dimensional shapes on the page. Sort the shapes by the number of vertices. Draw the shapes on the sorting mat. Use the words *alike* and *different* to tell how you sorted the shapes.

Name _____

alike	different

DIRECTIONS **2.** Place two-dimensional shapes on the page. Sort the shapes by the number of sides. Draw the shapes on the sorting mat. Use the words *alike* and *different* to tell how you sorted the shapes.

PROBLEM SOLVING

1

2

curve	no curve

DIRECTIONS **I.** I have a curve. What shape am I? Draw the shape. **2.** Draw to show shapes sorted by curves and no curves.

HOME ACTIVITY • Describe a shape and ask your child to name the shape that you are describing.

FOR MORE PRACTICE:
Standards Practice Book, pp. P191–P192

Name _____

Problem Solving • Draw to Join Shapes

Essential Question How can you solve problems using the strategy *draw a picture*?

🔑 Unlock the Problem

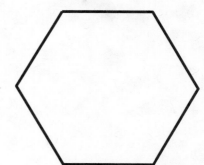

DIRECTIONS How can you join triangles to make the shapes? Draw and color the triangles.

Try Another Problem

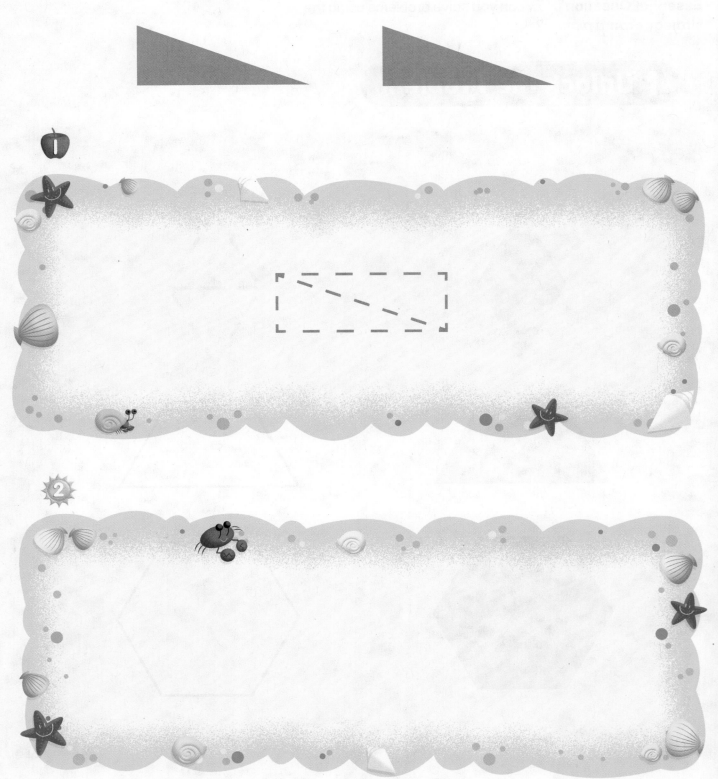

DIRECTIONS **1.** How can you join the two triangles to make a rectangle? Trace around the triangles to draw the rectangle. **2.** How can you join the two triangles to make a larger triangle? Use the triangle shapes to draw a larger triangle.

Name _____

Share and Show

3

4 ✓

DIRECTIONS   3. How can you join some of the squares to make a larger square?
Use the square shapes to draw a larger square.   **4.** How can you join some or all of
the squares to make a rectangle? Use the square shapes to draw a rectangle.

Chapter 9 • Lesson 12   four hundred three **403**

© Houghton Mifflin Harcourt Publishing Company

1

2

DIRECTIONS 1. Can you join these shapes to make a hexagon? Use the shapes to draw a hexagon. 2. Which shapes could you join to make the larger shape? Draw and color to show the shapes you used.

HOME ACTIVITY • Have your child join shapes to form a larger shape, and then tell you about the shape.

404 four hundred four

FOR MORE PRACTICE:
Standards Practice Book, pp. P193–P194

Chapter 9 Review/Test

Vocabulary

side

vertex

Concepts and Skills

2

3

4

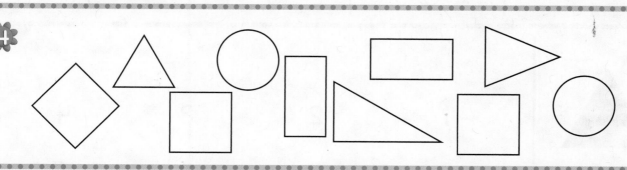

DIRECTIONS **1.** Draw a line from the word *side* to a side of each shape. Draw a line from the word *vertex* to a vertex, or corner, of each shape. (pp. 377, 385) **2.** Join triangles to make the shape. Draw and color the triangles you used. **3.** Join triangles and squares to make the shape. Draw and color the shapes you used. **4.** Use red to color the shapes with four vertices and four sides. Use green to color the shapes with curves. Use blue to color the shapes with three vertices and three sides.

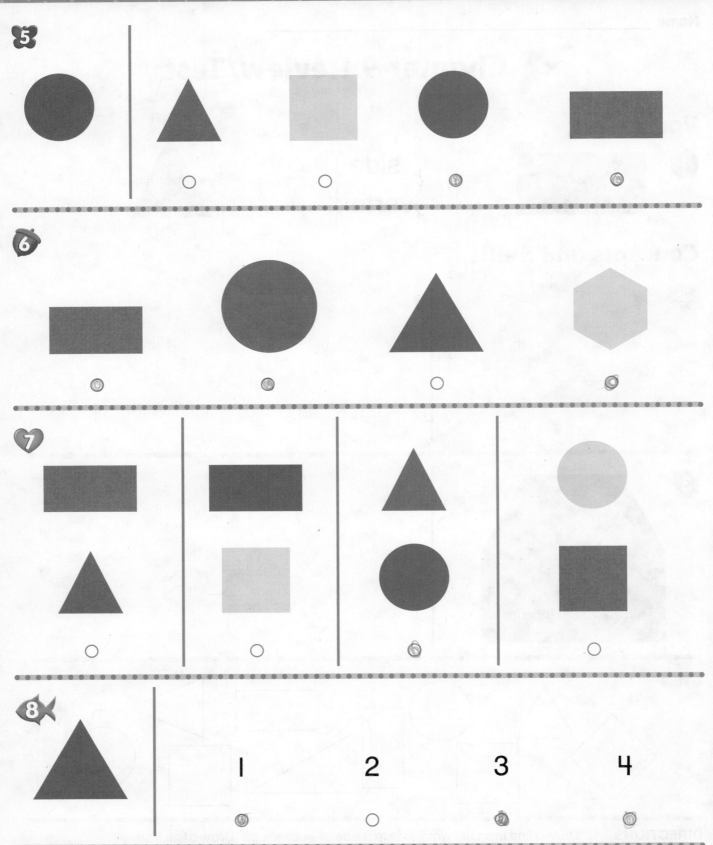

DIRECTIONS 5. Mark under the shape that matches the shape at the beginning of the row. **6.** Mark under the shape that is a hexagon. **7.** Mark under the set of shapes that both have four sides. **8.** Mark under the number that shows how many sides the triangle has.

Name _____

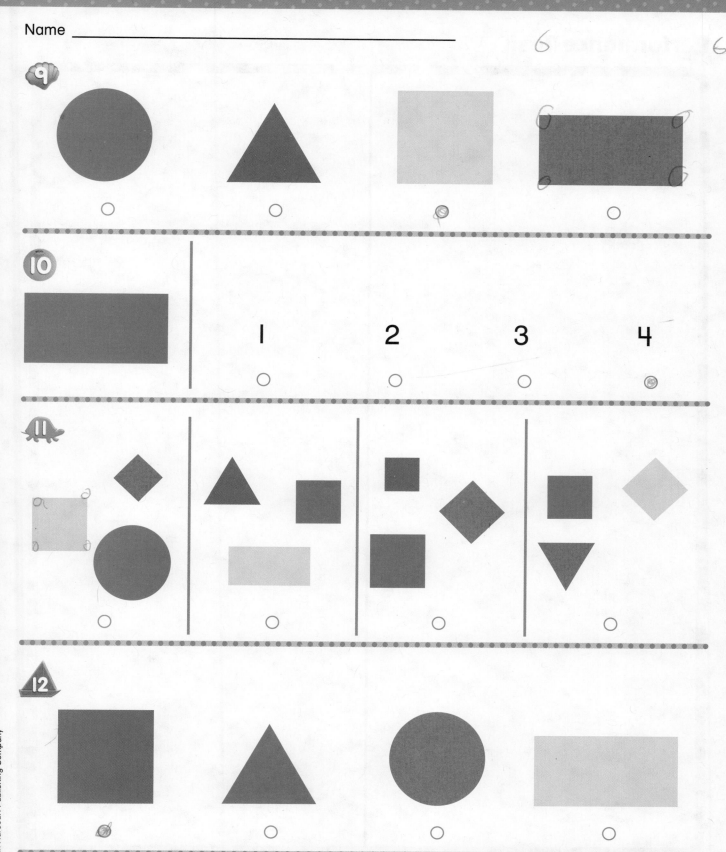

DIRECTIONS **9.** Mark under the shape that is a circle. **10.** Mark under the number that shows how many vertices the rectangle has **11.** Mark under the set that has all shapes with 4 sides. **12.** Mark under the shape that is a square.

Performance Task

PERFORMANCE TASK This task will assess the child's understanding of two-dimensional shapes.

Identify and Describe Three-Dimensional Shapes

Curious About Math with

Curious George

Many of the shapes in our environment are three-dimensional shapes.

Name some of the shapes you see in this picture.

Name _____

Show What You Know

Identify Shapes

 1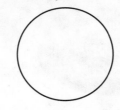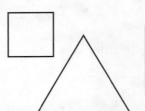

Describe Shapes

 2

_____ sides

_____ vertices

 3

_____ sides

_____ vertices

Sort Shapes

 4

DIRECTIONS **1.** Use red to color the squares. Use blue to color the triangles. **2–3.** Look at the shape. Write how many sides. Write how many vertices. **4.** Mark an X on the shapes with three sides.

FAMILY NOTE: This page checks your child's understanding of important skills needed for success in Chapter 10.

Name _____

Vocabulary Builder

circle

rectangle

square

triangle

DIRECTIONS Mark an X on the food shaped like a circle. Draw a line under the food shaped like a square. Circle the food shaped like a triangle.

GO Online
• eStudent Edition
• Multimedia eGlossary

Chapter 10

four hundred eleven **411**

Game Follow the Shapes

DIRECTIONS Choose a shape from START. Follow the path that has the same shapes. Draw a line to show the path to the END with the same shape.

Name _____

Three-Dimensional Shapes

Essential Question How can you show which shapes stack, roll, or slide?

Listen and Draw

does stack	does not stack

DIRECTIONS Place three-dimensional shapes on the page. Sort the shapes by whether they stack or do not stack. Describe the shapes. Match a picture of each shape to the shapes on the sorting mat. Glue the shape pictures on the sorting mat.

Chapter 10 · Lesson 1

four hundred thirteen **413**

Share and Show

1 ✓

roll

roll and stack

stack

DIRECTIONS 1. Place three-dimensional shapes on the page. Sort the shapes by whether they roll or stack. Describe the shapes. Match a picture of each shape to the shapes. Glue the shape pictures on the page.

414 four hundred fourteen

Name _Melanie_

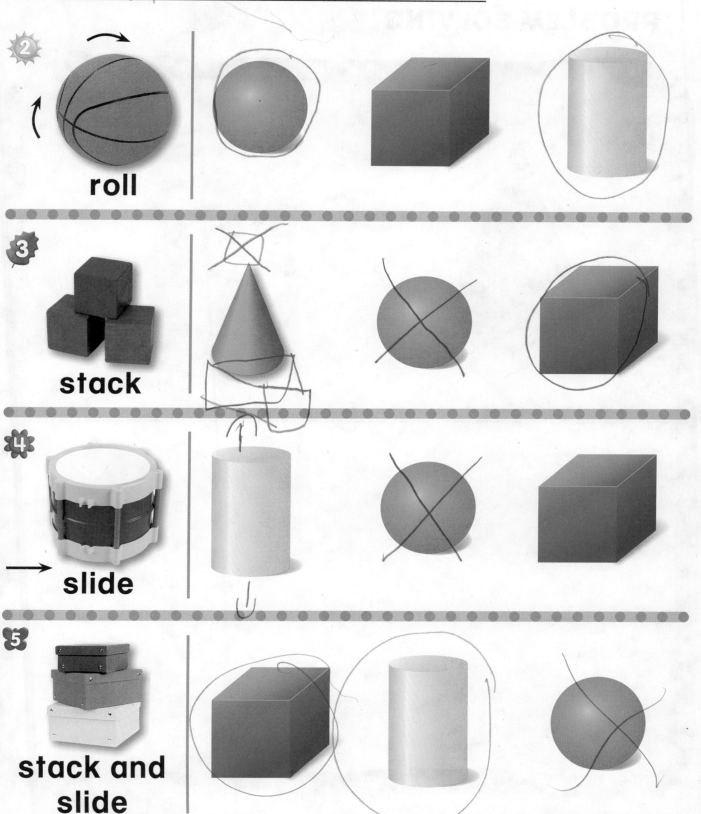

2 roll

3 stack

4 slide

5 stack and slide

DIRECTIONS **2.** Which shape does not roll? Mark an X on that shape. **3.** Which shapes do not stack? Mark an X on those shapes. **4.** Which shape does not slide? Mark an X on that shape. **5.** Which shape does not stack and slide? Mark an X on that shape.

PROBLEM SOLVING

1

2

DIRECTIONS 1. I roll and do not stack. Describe the shape. Mark an X on that shape. 2. Draw to show what you know about a real object that rolls and does not stack.

HOME ACTIVITY • Have your child identify and describe a household object that rolls and does not stack.

416 four hundred sixteen

ACTICE:
Book, pp. P199–P200

Name _____

Identify, Name, and Describe Spheres

Essential Question How can you identify, name, and describe spheres?

Listen and Draw  REAL WORLD

sphere	not a sphere

DIRECTIONS Place three-dimensional shapes on the page. Identify and name the sphere. Sort the shapes on the sorting mat. Describe the sphere. Match a picture of each shape to the shapes on the sorting mat. Glue the shape pictures on the sorting mat.

Chapter 10 • Lesson 2

1

sphere

flat surface

curved surface

2 ✓

   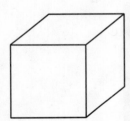

 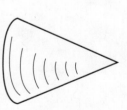 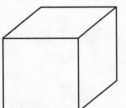 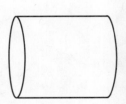

DIRECTIONS 1. Look at the sphere. Circle the words that describe a sphere. **2.** Color the spheres.

DIRECTIONS **3.** Identify the objects that are shaped like a sphere. Mark an X on those objects.

Chapter 10 • Lesson 2

four hundred nineteen **419**

PROBLEM SOLVING REAL WORLD

1.

2.

DIRECTIONS 1. I have a curved surface. Which shape am I? Mark an X on that shape. **2.** Draw to show what you know about a real object that is shaped like a sphere.

HOME ACTIVITY • Have your child identify and describe a household object that is shaped like a sphere.

FOR MORE PRACTICE:
Standards Practice Book, pp. P201–P202

Name _____

Identify, Name, and Describe Cubes

Essential Question How can you identify, name, and describe cubes?

Listen and Draw REAL WORLD

cube	not a cube

DIRECTIONS Place three-dimensional shapes on the page. Identify and name the cube. Sort the shapes on the sorting mat. Describe the cube. Match a picture of each shape to the shapes on the sorting mat. Glue the shape pictures on the sorting mat.

Chapter 10 • Lesson 3

cube

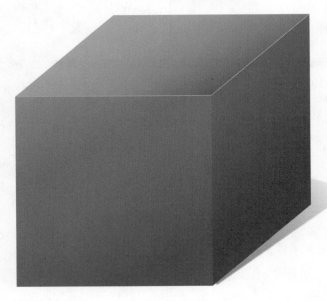

flat surface

curved surface

_____

_ _ _ _ _ _ _

_____ **flat surfaces**

DIRECTIONS 1. Look at the cube. Circle the words that describe a cube. 2. Use a cube to count how many flat surfaces. Write the number.

Name _____

DIRECTIONS **3.** Identify the objects that are shaped like a cube. Mark an X on those objects.

Chapter 10 • Lesson 3

four hundred twenty-three **423**

PROBLEM SOLVING *REAL* **WORLD**

1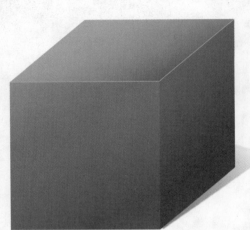

2

DIRECTIONS 1. I have 6 flat surfaces. Which shape am I? Mark an X on that shape. 2. Draw to show what you know about a real object that is shaped like a cube.

 HOME ACTIVITY • Have your child identify and describe a household object that is shaped like a cube.

424 four hundred twenty-four

FOR MORE PRACTICE: Standards Practice Book, pp. P203–P204

Identify, Name, and Describe Cylinders

Essential Question How can you identify, name, and describe cylinders?

Listen and Draw REAL WORLD

cylinder	not a cylinder

DIRECTIONS Place three-dimensional shapes on the page. Identify and name the cylinder. Sort the shapes on the sorting mat. Describe the cylinder. Match a picture of each shape to the shapes on the sorting mat. Glue the shape pictures on the sorting mat.

1

cylinder

flat surface

curved surface

---

 2 ✓

_____

— — — — —

_____ **flat surfaces**

DIRECTIONS **1.** Look at the cylinder. Circle the words that describe a cylinder. **2.** Use a cylinder to count how many flat surfaces. Write the number.

Name _____

DIRECTIONS 3. Identify the objects that are shaped like a cylinder. Mark an X on those objects.

Chapter 10 · Lesson 4

four hundred twenty-seven **427**

PROBLEM SOLVING

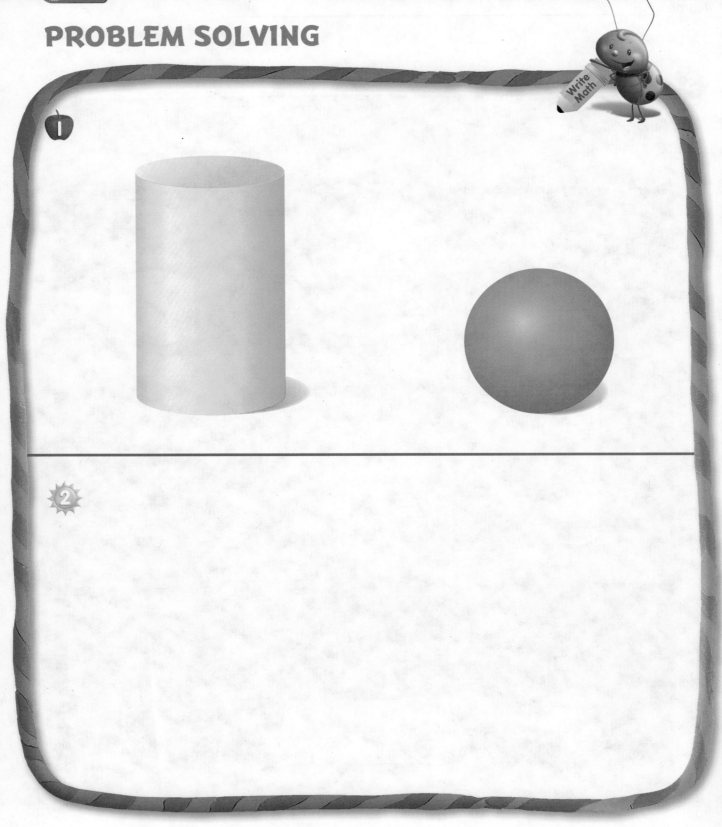

DIRECTIONS **1.** I have 2 flat surfaces. Which shape am I? Mark an X on that shape. **2.** Draw to show what you know about a real object that is shaped like a cylinder.

HOME ACTIVITY • Have your child identify and describe a household object that is shaped like a cylinder.

FOR MORE PRACTICE: Standards Practice Book, pp. P205–P206

Name _____

Identify, Name, and Describe Cones

Essential Question How can you identify, name, and describe cones?

Listen and Draw

cone	not a cone

DIRECTIONS Place three-dimensional shapes on the page. Identify and name the cone. Sort the shapes on the sorting mat. Describe the cone. Match a picture of each shape to the shapes on the sorting mat. Glue the shape pictures on the sorting mat.

Chapter 10 · Lesson 5

four hundred twenty-nine **429**

cone

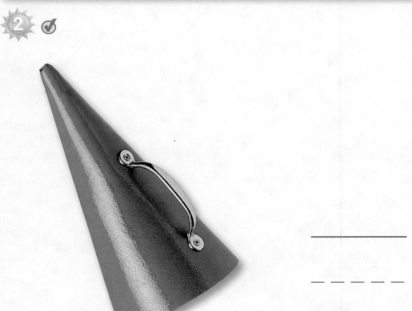

flat surface

curved surface

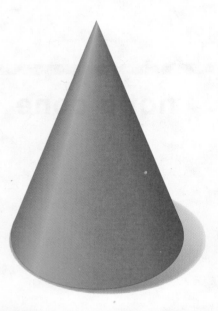

_____

_ _ _ _ _

_____ **flat surface**

DIRECTIONS 1. Look at the cone. Circle the words that describe a cone. 2. Use a cone to count how many flat surfaces. Write the number.

Name _____

DIRECTIONS   3. Identify the objects that are shaped like a cone. Mark an X on those objects.

HOME ACTIVITY • Have your child identify and describe a household object that is shaped like a cone.

Chapter 10 • Lesson 5

FOR MORE PRACTICE: Standards Practice Book, pp. P207–P208

✓ Mid-Chapter Checkpoint

Concepts and Skills

 1

2 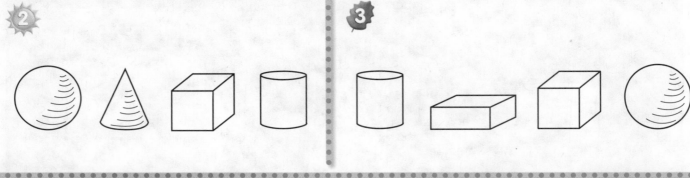 **3**

4

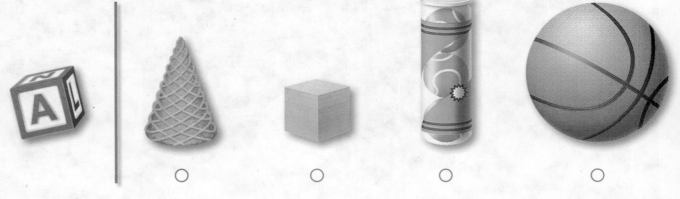

DIRECTIONS 1. Mark an X on the object that is shaped like a cylinder.
2. Color the sphere.   3. Color the cube.   4. Mark under the
object that has the same shape as the shape at the beginning of the row.

432 four hundred thirty-two

Name _____

Problem Solving • Two- and Three-Dimensional Shapes

Essential Question How can you solve problems using the strategy *use logical reasoning*?

  **Unlock the Problem** REAL WORLD

two-dimensional shapes	three-dimensional shapes

DIRECTIONS Place shapes on the page. Sort the shapes on the sorting mat into sets of two-dimensional and three-dimensional shapes. Match a picture of each shape to a shape on the sorting mat. Glue the shape pictures on the sorting mat.

Chapter 10 • Lesson 6

Try Another Problem

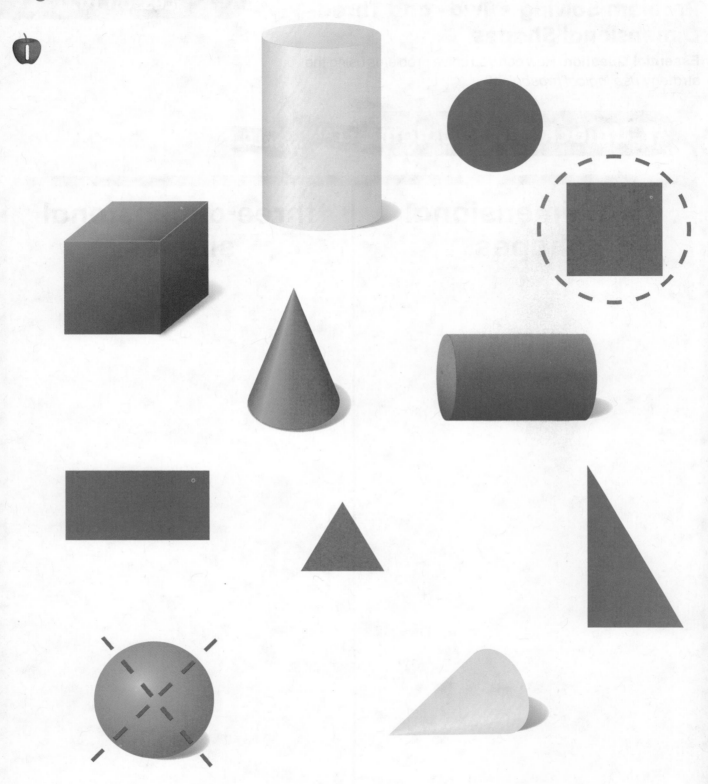

DIRECTIONS 1. Identify the two-dimensional or flat shapes. Trace the circle around the square. Circle the other flat shapes. Identify the three-dimensional or solid shapes. Trace the X on the sphere. Mark an X on the other solid shapes.

434 four hundred thirty-four

Name _____

Share and Show

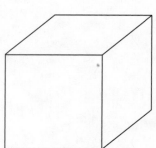

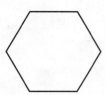
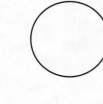

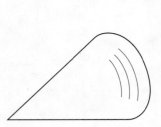

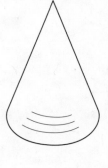

DIRECTIONS 2. Identify the two-dimensional or flat shapes. Use red to color the flat shapes. Identify the three-dimensional or solid shapes. Use blue to color the solid shapes.

Chapter 10 · Lesson 6 four hundred thirty-five **435**

On Your Own

1

2

© Houghton Mifflin Harcourt Publishing Company

DIRECTIONS **1.** Draw to show what you know about a flat shape. Name the shape. **2.** Draw to show what you know about a solid shape. Name the shape.

HOME ACTIVITY • Have your child identify a household object that is shaped like a three-dimensional shape. Have him or her name the three-dimensional shape.

FOR MORE PRACTICE:
Standards Practice Book, pp. P209–P210

Name _____

Above and Below

Essential Question How can you use the terms *above* and *below*, to describe shapes in the environment?

Listen and Draw  REAL WORLD

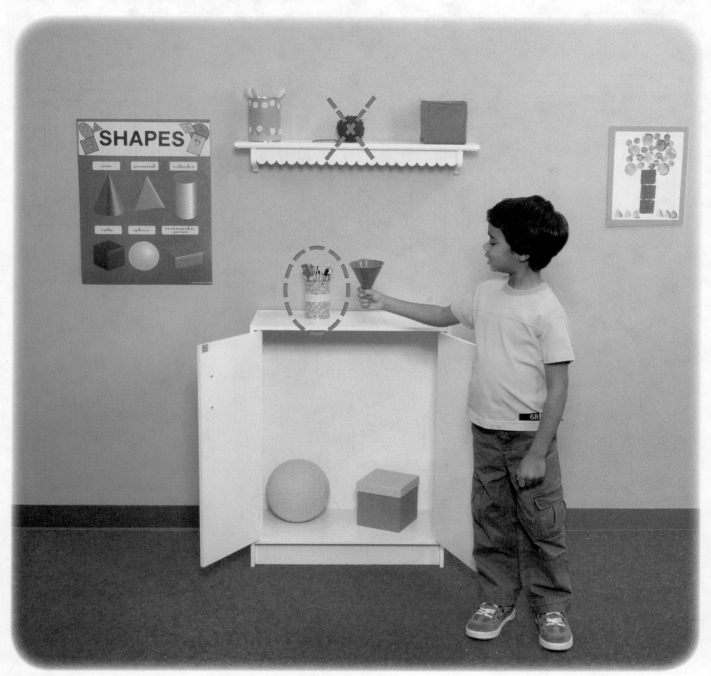

DIRECTIONS Trace the circle around the object shaped like a cylinder that is below the shelf. Trace the X on the object shaped like a sphere that is above the cabinet.

Chapter 10 • Lesson 7

1

DIRECTIONS 1. Circle the object that is shaped like a cone below the play set. Mark an X on the object that is shaped like a cube above the play set. Color the object that is shaped like a cylinder above the play set.

Name _____

DIRECTIONS   2. Circle the ball that is above the net. Mark an X on the
box that is directly below the net.

PROBLEM SOLVING · REAL WORLD

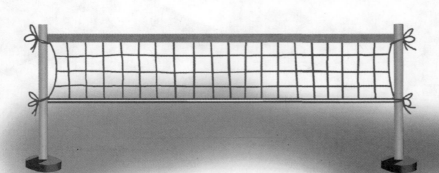

DIRECTIONS Draw to show what you know about real world three-dimensional objects that might be above or below the net. Tell a friend about your drawing as you name the shape of the objects.

HOME ACTIVITY • Tell your child you are thinking of something in the room that is above or below another object. Have your child tell you what the object might be.

FOR MORE PRACTICE:
Standards Practice Book, pp. P211–P212

Name _____

Beside and Next To

Essential Question How can you use the terms *beside* and *next to* to describe shapes in the environment.

Listen and Draw  REAL WORLD

DIRECTIONS Trace the X on the object shaped like a cone that is beside the object shaped like a sphere. Trace the circle on the object shaped like a sphere that is next to the object shaped like a cube.

Chapter 10 • Lesson 8

Share and Show

DIRECTIONS 1. Mark an X on the bead shaped like a cube that is beside the bead shaped like a cone. Draw a circle around the bead shaped like a cone that is next to the bead shaped like a cylinder. Use the words *next to* and *beside* to name the position of other bead shapes.

Name _____

DIRECTIONS **2.** Mark an X on the object shaped like a cylinder that is
next to the object shaped like a sphere. Draw a circle around the object
shaped like a cone that is beside the object shaped like a cube. Use the
words *next to* and *beside* to describe the position of other package shapes.

PROBLEM SOLVING REAL WORLD

DIRECTIONS Draw or use pictures to show what you know about real world three-dimensional objects beside and next to other objects.

HOME ACTIVITY • Tell your child you are thinking of something in the room that is beside or next to another object. Have your child tell you the shape of the object.

444 four hundred forty-four

FOR MORE PRACTICE:
Standards Practice Book, pp. P213–P214

Name _____

In Front Of and Behind

Essential Question How can you use the terms *in front of* and *behind* to describe shapes in the environment.

Listen and Draw · REAL WORLD

DIRECTIONS Trace the X on the object shaped like a sphere that is in front of the object shaped like a cube. Trace the circle around the object shaped like a cylinder that is behind the object shaped like a cube.

Chapter 10 • Lesson 9

Share and Show

DIRECTIONS **I.** Mark an X on the object shaped like a cylinder that is behind the object shaped like a cube. Draw a circle around the object shaped like a sphere that is in front of the object shaped like a cone. Use the words *in front of* and *behind* to name the position of other shapes.

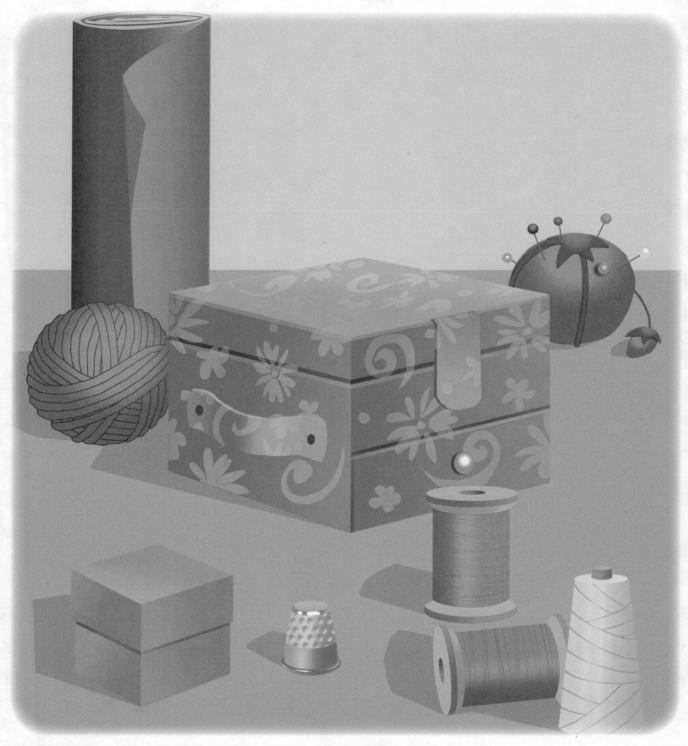

DIRECTIONS 2. Mark an X on the object shaped like a cube that is in front of the object shaped like a cylinder. Draw a circle around the object shaped like a cylinder that is behind the object shaped like a sphere. Use the words *in front of* and *behind* to name the position of other shaped objects.

Chapter 10 • Lesson 9   four hundred forty-seven **447**

PROBLEM SOLVING REAL WORLD

DIRECTIONS Draw or use pictures to show what you know about real world three-dimensional objects in front of and behind other objects.

HOME ACTIVITY • Tell your child you are thinking of something in the room that is in front of or behind another object. Have your child tell you the shape of the object.

FOR MORE PRACTICE:
Standards Practice Book, pp. P215–P216

Name _____

Vocabulary

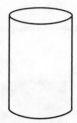  

Concepts and Skills

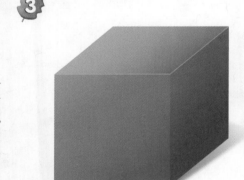

flat surface

curved surface

DIRECTIONS **1.** Use blue to color the sphere. Use green to color the cylinder. Use red to color the cube. Use purple to color the cone. (pp. 417, 421, 425, 429) **2.** Mark an X on the object that rolls and does not stack. **3.** Circle the words that describe a cube.

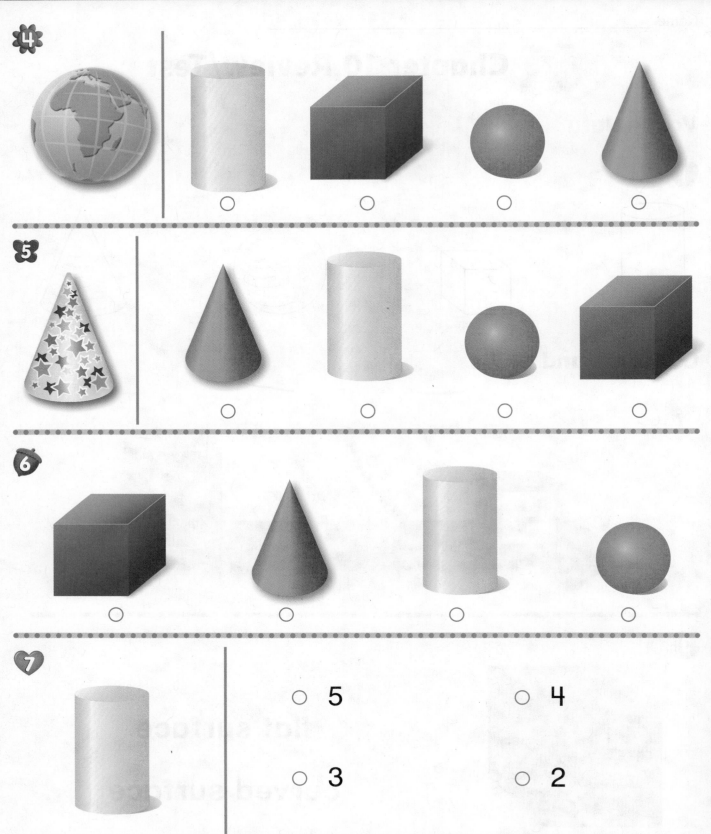

4 ○ ○ ○ ○

5 ○ ○ ○ ○

6 ○ ○ ○ ○

7 ○ 5   ○ 4

○ 3   ○ 2

DIRECTIONS   4–5. Mark under the shape that is the same shape as the object at the beginning of the row.   **6.** Mark under the cube.   **7.** Mark beside the number that shows how many flat surfaces the cylinder has.

450 four hundred fifty

Name _____

8

○   ○   ○   ○

9

○   ○   ○   ○

10

○   ○   ○   ○

11   ○   ○   ○   ○

DIRECTIONS **8.** Mark under the shape that is flat. **9.** Mark under the set that shows a cube above a cylinder. **10.** Mark under the set that shows a cylinder beside a cube **11.** Mark above the shape that is behind the cube in the truck.

Chapter 10   four hundred fifty-one **451**

Performance Task

PERFORMANCE TASK   This task will assess the child's understanding of three-dimensional shapes.

Plants all Around

written by Tami Morton

Representing, relating, and operating on whole
numbers, initially with sets of objects

Two leaves fall from a tree.

Circle the leaf that is longer.

Science

Why do plants have leaves?

Two flowers grow near a wall.

Circle the flower that is shorter.

Science

Why do plants have flowers?

These carrots grow under the ground.

Circle the carrot that is longer.

Science

Why do plants have roots?

Cattails can be short or tall.

Circle the two cattails that are about the same height.

Science

Why do plants have stems?

One leaf is shorter than the other leaf.

Draw a leaf that is about the same length as the shorter leaf.

Science

How are all these plants the same?

Write About the Story

Draw a purple flower. Make it shorter than the orange flower and taller than the yellow flower.

Vocabulary Review

| longer | taller |
| shorter | same |

Longer and Shorter

1. Look at the carrot. Draw a shorter carrot on the left.
Draw a longer carrot on the right.

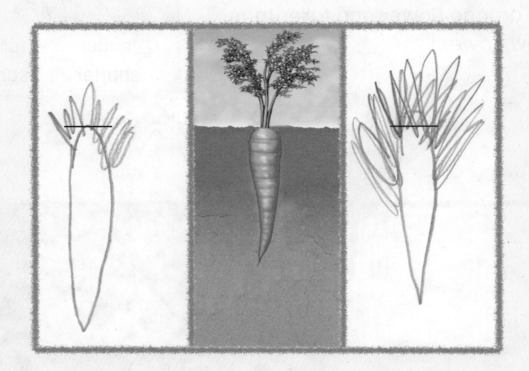

2. Look at the leaf.
Draw a longer leaf above it.
Draw a shorter leaf below it.

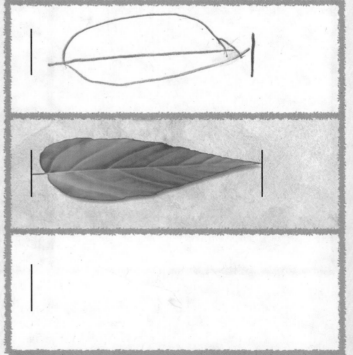

Curious About Math with

Curious George

A playground is an area designed for children to play.

- Which person on the park bench is bigger?

Name _____

Show What You Know ✓

More and Fewer

_____   _____

- - - - - - - -   - - - - - - - -

_____   _____

 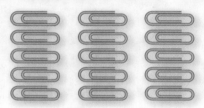   

_____   _____

- - - - - - - -   - - - - - - - -

_____   _____

Compare Numbers

_____

- - - - - - - -

_____

_____

_____

_____

DIRECTIONS **1.** Write how many in each set. Circle the set with fewer objects. **2.** Write how many in each set. Circle the set with more objects. **3.** Write how many cubes in each set. Circle the greater number.

FAMILY NOTE: This page checks your child's understanding of important skills needed for success in Chapter 11.

© Houghton Mifflin Harcourt Publishing Company

GO Online   Assessment Options
Soar to Success Math

462 four hundred sixty-two

Name _____

Vocabulary Builder

bigger

smaller

DIRECTIONS Are there more flowers in the bigger pot or the smaller pot? Circle to show the pot with more flowers.

GO
Online
• eStudent Edition
• Multimedia eGlossary

Game

Connecting Cube Challenge

START

END

© Houghton Mifflin Harcourt Publishing Company

DIRECTIONS Take turns with a partner tossing the number cube. Move your marker that number of spaces. If a player lands on a cube, he or she takes a cube for making a cube train. At the end of the game, players compare cube trains. Each player identifies the number of cubes in his or her cube train. If one player has a greater number of cubes, partners should identify that as the larger quantity of cubes.

MATERIALS game markers, number cube (1–6), connecting cubes

464 four hundred sixty-four

Name _____

Compare Lengths

Essential Question How can you compare the lengths of two objects?

Listen and Draw  REAL WORLD

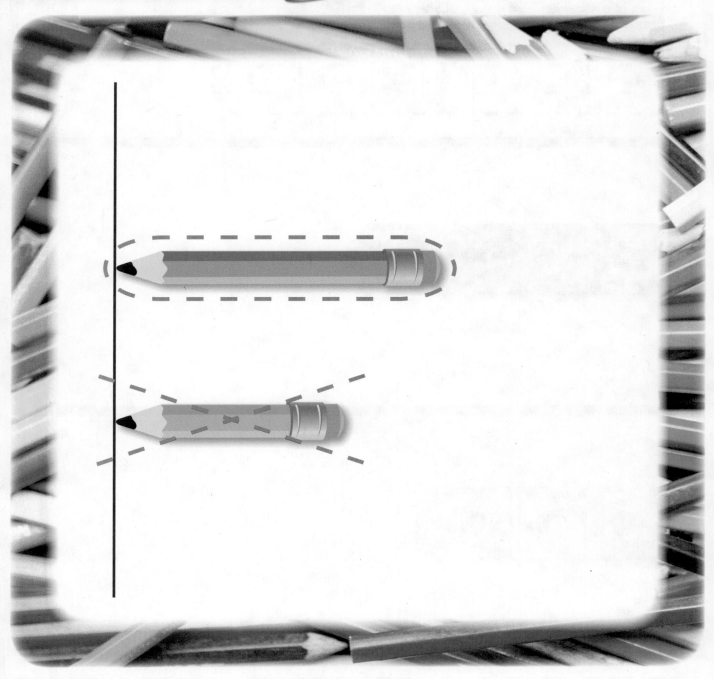

DIRECTIONS Look at the pencils. Compare the lengths of the two pencils. Use the words *longer than*, *shorter than*, or *about the same length* to describe the lengths. Trace around the longer pencil. Trace the X on the shorter pencil.

Chapter 11 • Lesson 1

Share and Show

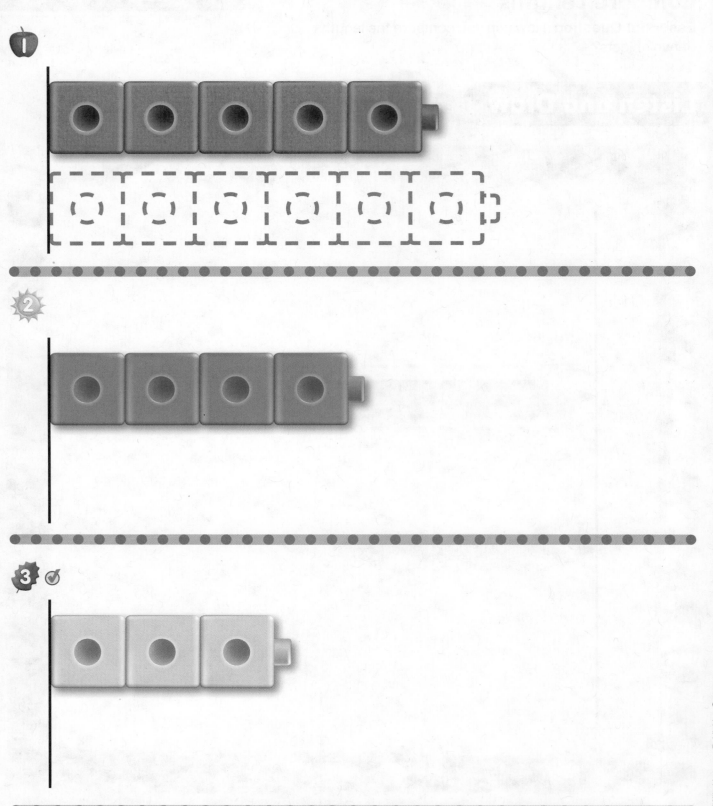

1

2

3 ✓

DIRECTIONS **I.** Place cubes on the longer cube train. Trace and color the cube train. **2–3.** Make a cube train that is longer than the cube train shown. Draw and color the cube train.

466 four hundred sixty-six

© Houghton Mifflin Harcourt Publishing Company

4 ✓

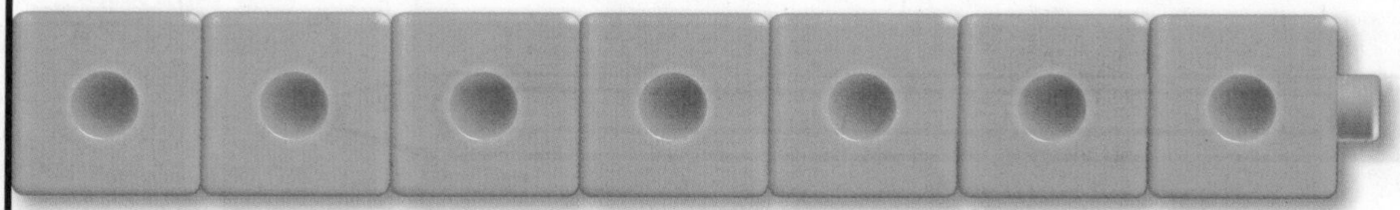

5

6

DIRECTIONS **4–6.** Make a cube train that is shorter than the cube train shown. Draw and color the cube train.

PROBLEM SOLVING REAL WORLD

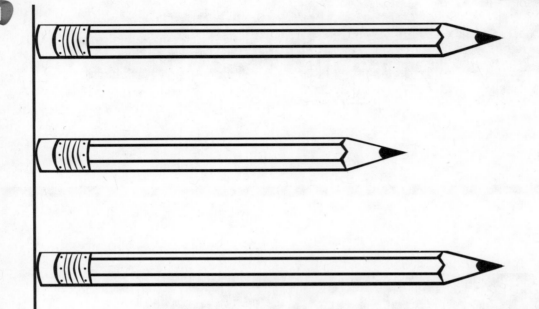

DIRECTIONS 1. Two of these pencils are about the same length. Color those pencils. 2. Draw to show what you know about two objects that are about the same length. Tell a friend about your drawing.

HOME ACTIVITY • Show your child a pencil and ask him or her to find an object that is longer than the pencil. Repeat with an object that is shorter than the pencil.

468 four hundred sixty-eight

Name _____

Compare Heights

Essential Question How can you compare
the heights of two objects?

Listen and Draw REAL WORLD

DIRECTIONS Look at the chairs. Compare the heights of the two chairs.
Use the words *taller than*, *shorter than*, or *about the same height* to describe
the heights. Trace around the taller chair. Trace the X on the shorter chair.

Share and Show

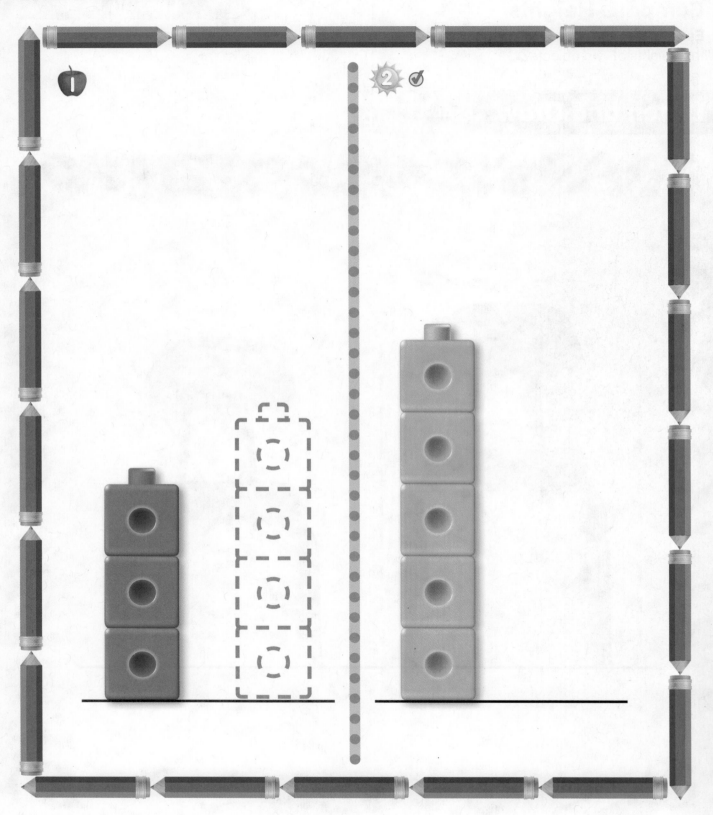

DIRECTIONS 1. Place cubes on the taller cube tower. Trace and color the cube tower. 2. Make a cube tower that is taller than the cube tower shown. Draw and color the cube tower.

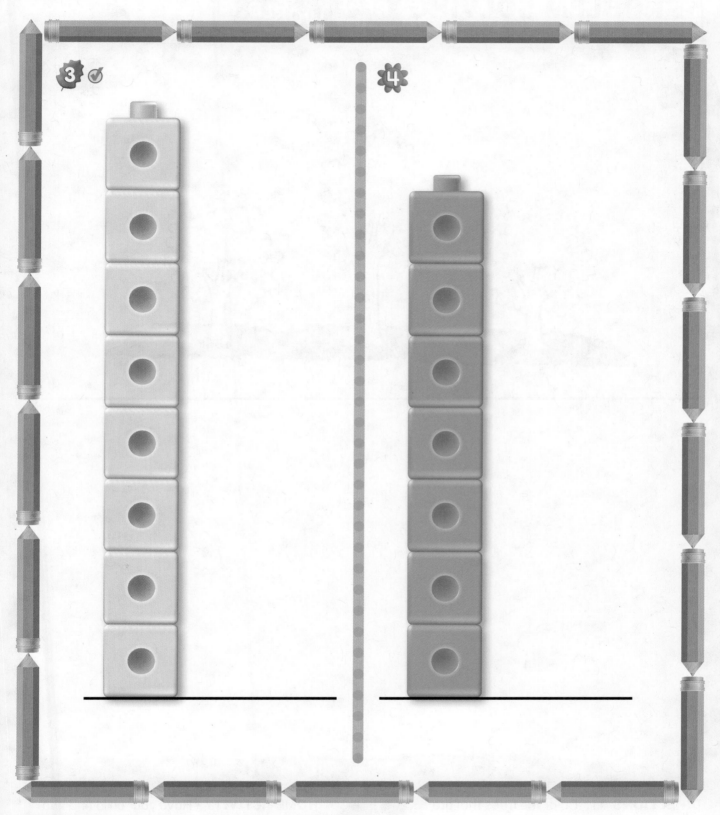

DIRECTIONS 3–4. Make a cube tower that is shorter than the cube tower shown. Draw and color the cube tower.

Chapter 11 · Lesson 2

four hundred seventy-one **471**

PROBLEM SOLVING REAL WORLD

1

2

_____   _____

DIRECTIONS **1.** Color the trees that are about the same height. **2.** Draw to show what you know about two cube towers that are about the same height. Tell a friend about your drawing.

HOME ACTIVITY • Have your child find two objects such as plastic toys or stuffed animals. Have him or her place the objects side by side to compare the heights. Ask your child which object is taller and which object is shorter.

FOR MORE PRACTICE:
Standards Practice Book, pp. P223–P224

Name _____

Problem Solving • Direct Comparison

Essential Question How can you solve problems using the strategy *draw a picture*?

 **Unlock the Problem** REAL WORLD

DIRECTIONS Compare the lengths or heights of two classroom objects. Draw the objects. Tell a friend about your drawing.

Try Another Problem

DIRECTIONS **1.** Find two small classroom objects. Place one end of each object on the line. Compare the lengths. Draw the objects. Say *longer than*, *shorter than*, or *about the same length* to describe the lengths. Circle the longer object.

Name _____

Share and Show

DIRECTIONS **2.** Find two small classroom objects. Place one end of each object on the line. Compare the heights. Draw the objects. Say *taller than*, *shorter than*, or *about the same height* to describe the heights. Circle the shorter object.

HOME ACTIVITY • Show your child two objects of different lengths. Have him or her align the ends of both objects to compare the lengths and tell which object is shorter and which object is longer.

Chapter 11 • Lesson 3 **FOR MORE PRACTICE:** Standards Practice Book, pp. P225–P226

four hundred seventy-five **475**

Concepts and Skills

1

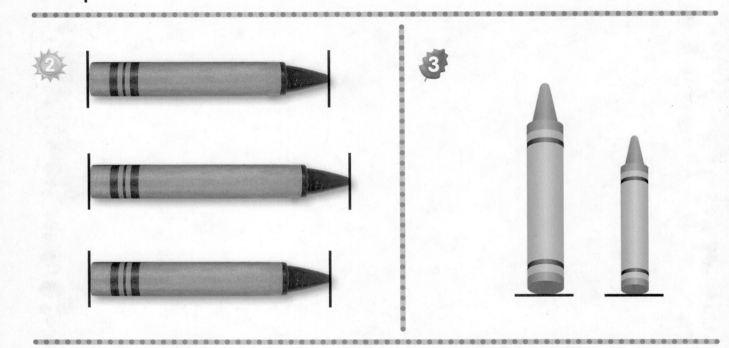

2 **3**

4

○   ○   ○   ○

DIRECTIONS **1.** Make a cube train that is shorter than the one shown. Draw the cube train.   **2.** Circle the crayons that are about the same length.   **3.** Circle the crayon that is shorter.   **4.** Find the set with two pencils that are about the same length. Mark under the set.

Name _____

Compare Weights

Essential Question How can you compare the weights of two objects?

Listen and Draw REAL WORLD

DIRECTIONS Look at the picture. Compare the weights of the two objects. Use the words *heavier than*, *lighter than*, or *about the same weight* to describe the weights. Trace the circle around the lighter object. Trace the X on the heavier object.

Chapter 11 • Lesson 4

Share and Show

 left    right

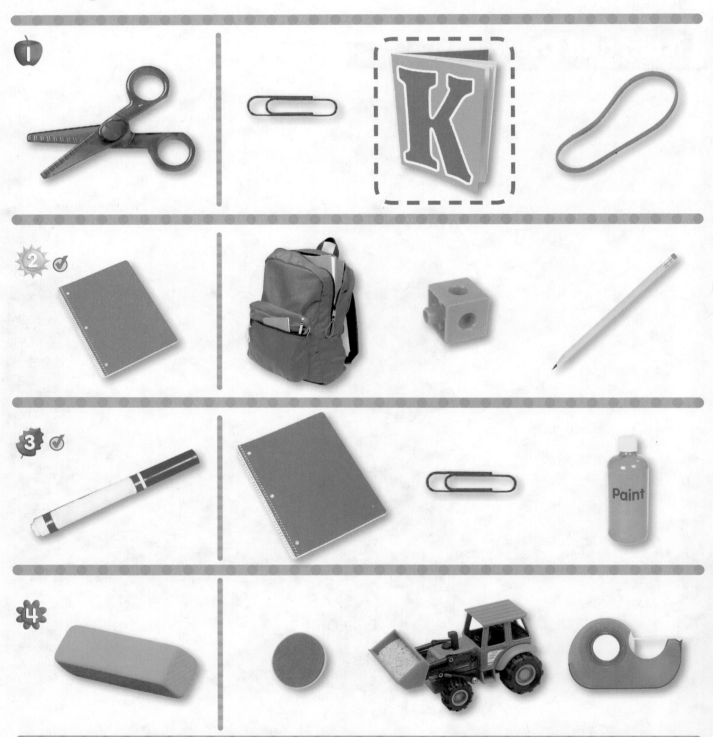

DIRECTIONS Find the first object in the row, and hold it in your left hand. Find the rest of the objects in the row, and hold each object in your right hand.   **1.** Trace around the object that is heavier than the object in your left hand.   **2.** Circle the object that is heavier than the object in your left hand.   **3–4.** Circle the object that is lighter than the object in your left hand.

478 four hundred seventy-eight

© Houghton Mifflin Harcourt Publishing Company

Name _____

5

6

DIRECTIONS Find a book in the classroom. **5.** Find a classroom object that is lighter than the book. Draw it in the work space. **6.** Find a classroom object that is heavier than the book. Draw it in the work space.

Chapter 11 · Lesson 4

four hundred seventy-nine **479**

PROBLEM SOLVING REAL WORLD

DIRECTIONS Draw to show what you know about comparing the weights of two objects. Tell a friend about your drawing.

HOME ACTIVITY • Have your child compare the weights of two household objects. Then have him or her use the terms *heavier* and *lighter* to describe the weights.

FOR MORE PRACTICE:
Standards Practice Book, pp. P227–P228

Length, Height, and Weight

Essential Question How can you describe several ways to measure one object?

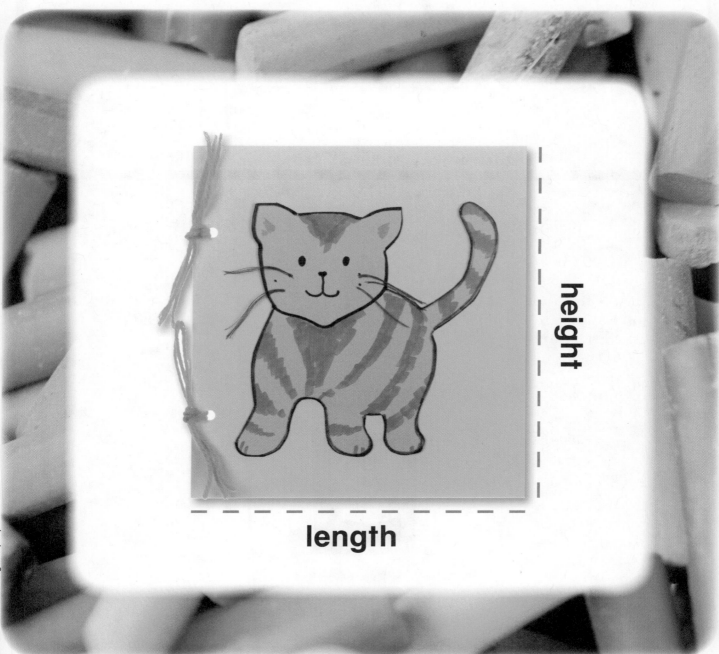

height

length

DIRECTIONS Look at the book. Trace your finger over the line that shows how to measure the height of the book. Trace your finger over the line that shows how to measure the length of the book. Talk about another way to measure the book.

Share and Show

DIRECTIONS   1–2. Use red to trace the line that shows how to measure the length. Use blue to trace the line that shows how to measure the height. Talk about another way to measure the object.

Name _____

 3

 4

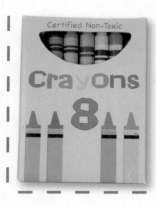

 5

6

DIRECTIONS **3–6.** Use red to trace the line that shows how to measure the length. Use blue to trace the line that shows how to measure the height. Talk about another way to measure the object.

PROBLEM SOLVING REAL WORLD

Write Math

DIRECTIONS   Draw to show what you know about measuring an object in more than one way.

HOME ACTIVITY • Show your child a household object that can be easily measured by length, height, and weight. Ask him or her to describe the different ways to measure the object.

484   four hundred eighty-four

✓ 🏴 Chapter 11 Review/Test

Vocabulary

 1

taller   shorter

 **2**

heavier   lighter

Concepts and Skills

 3

 4

DIRECTIONS **1–2.** Draw lines to match the words to the objects.
(pp. 469–472, 477–480) **3.** This is shorter than another tree. Draw to show the other tree. **4.** Look at the objects. Mark an X on the lighter object. Circle the heavier object.

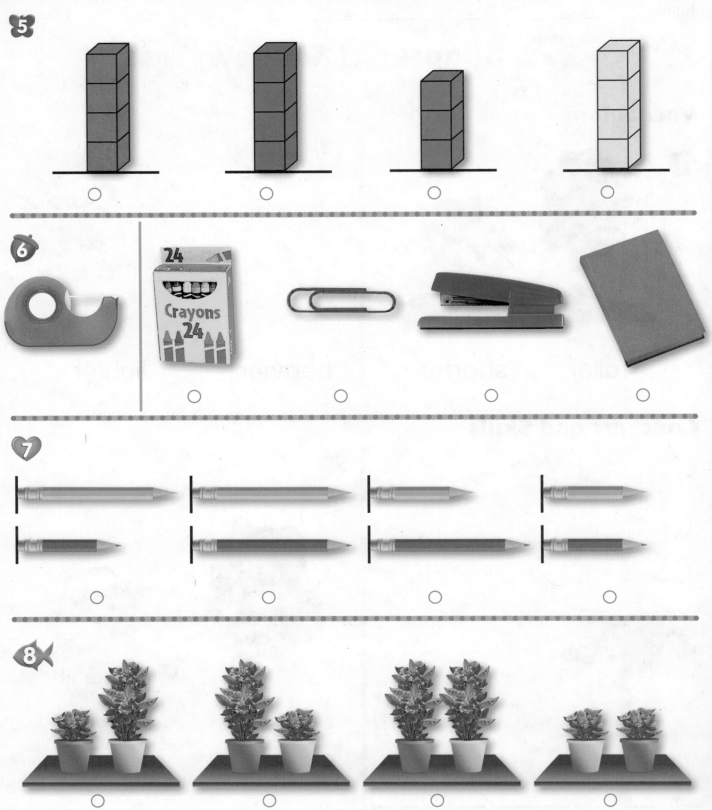

DIRECTIONS 5. Mark under the block tower that is shorter than the other towers.
6. Mark under the object that is lighter than the object at the beginning of the row.
7. Mark under the set that shows that the orange pencil is shorter than the green pencil. **8.** Mark under the picture that shows that the plant in the yellow pot is taller than the plant in the blue pot.

Name _____

9

○   ○   ○   ○

10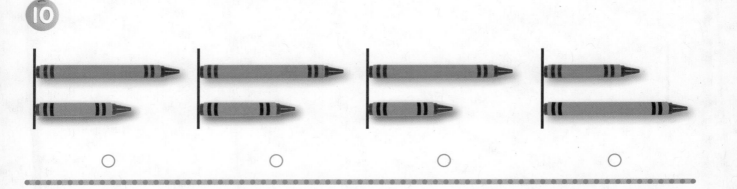

○   ○   ○   ○

11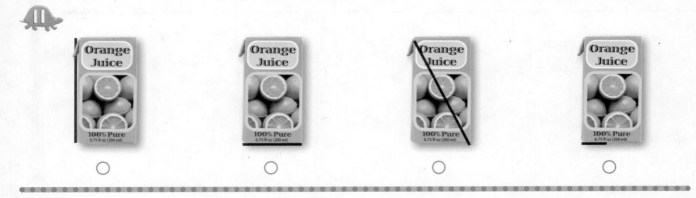

○   ○   ○   ○

12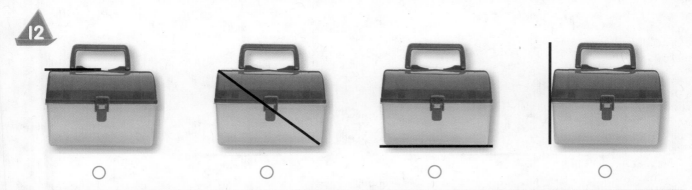

○   ○   ○   ○

DIRECTIONS **9.** Mark under the object that is heavier than the book at the beginning of the row. **10.** Mark under the set that shows the green crayon is longer than the blue crayon. **11.** Mark under the juicebox that has a line that shows how to measure height. **12.** Mark under the lunchbox that has a line that shows how to measure length.

Performance Task

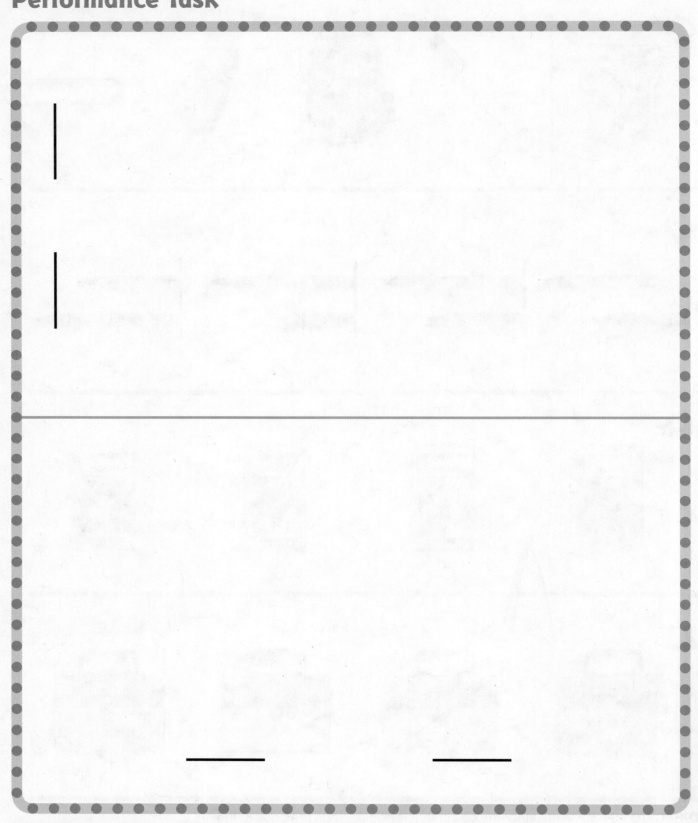

PERFORMANCE TASK This task will assess the child's understanding of measurement.

488 four hundred eighty-eight

Chapter 12
Classify and Sort Data

Curious About Math with
Curious George

Primary colors are blue, red, and yellow.

• How many primary colors is the girl sorting?

Name _____

Color and Shape

 1

2

Compare Sets

3  _____

 _____

4

DIRECTIONS **1.** Circle the fruits that are red. **2.** Circle the triangles. **3.** Count and write how many in each set. Circle the set with more objects. **4.** Count and write how many in each set. Circle the set with fewer objects.

FAMILY NOTE: This page checks your child's understanding of important skills needed for success in Chapter 12.

GO Online Assessment Options
Soar to Success Math

Name _____

different

alike

DIRECTIONS Tell what you know about the ladybugs. Some of the ladybugs are different. Circle those ladybugs and tell why they are different. Tell what you know about the butterflies.

GO Online
• eStudent Edition
• Multimedia eGlossary

Chapter 12

Game

At the Farm

© Houghton Mifflin Harcourt Publishing Company

DIRECTIONS Use the picture to play I Spy with a partner. Decide who will go first. Player 1 looks at the picture, selects an object, and tells Player 2 the color of the object. Player 2 must guess what Player 1 sees. Once Player 2 guesses correctly, it is his or her turn to choose an object and have Player 1 guess.

492 four hundred ninety-two

Name _____

Algebra • Classify and Count by Color

Essential Question How can you classify and count objects by color?

Listen and Draw

not

DIRECTIONS Choose a color. Use that color crayon to color the clouds. Sort and classify a handful of shapes into a set of that color and a set of not that color. Draw and color the shapes.

Share and Show

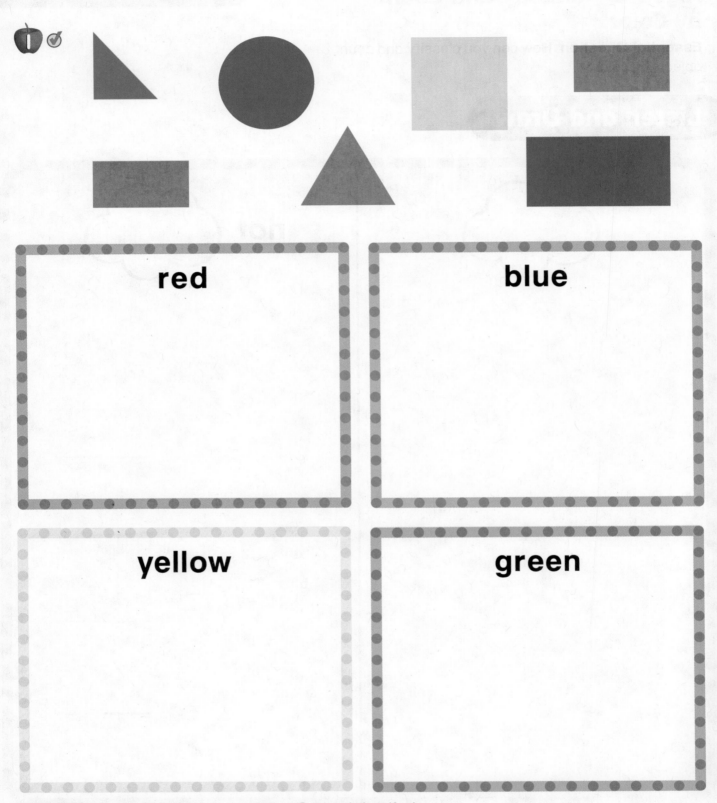

red

blue

yellow

green

DIRECTIONS 1. Place shapes as shown. Sort and classify the shapes by the category of color. Draw and color the shapes in each category.

© Houghton Mifflin Harcourt Publishing Company

494 four hundred ninety-four

Name _____

1

red	blue
yellow	green

_____
- - - - - - - -
_____

③

2

red	blue
yellow	green

_____
- - - - - - - -
_____

④

3

red	blue
yellow	green

_____
- - - - - - - -
_____

DIRECTIONS Look at the categories in Exercise 1. Count how many in each category. **2.** Circle the categories that have one shape. Write the number. **3.** Circle the category that has two shapes. Write the number. **4.** Circle the category that has 3 shapes. Write the number.

PROBLEM SOLVING

DIRECTIONS **1.** How are these shapes sorted and classified? Draw one more shape in each category. **2.** Draw to show what you know about sorting and classifying by color.

HOME ACTIVITY • Provide your child with different colors of the same objects, such as straws, socks, or marbles. Ask him or her to sort and classify the objects into two sets, a set of all one color and a set of all the other colors.

496 four hundred ninety-six

FOR MORE PRACTICE:
Standards Practice Book, pp. P235–P236

Name _____

Algebra • Classify and Count by Shape

Essential Question How can you classify and count objects by shape?

Listen and Draw

	not

DIRECTIONS Choose a shape. Draw the shape at the top of each side. Sort and classify a handful of shapes into a set of the shape you chose and a set not that shape. Draw and color the shapes.

Chapter 12 • Lesson 2

Share and Show

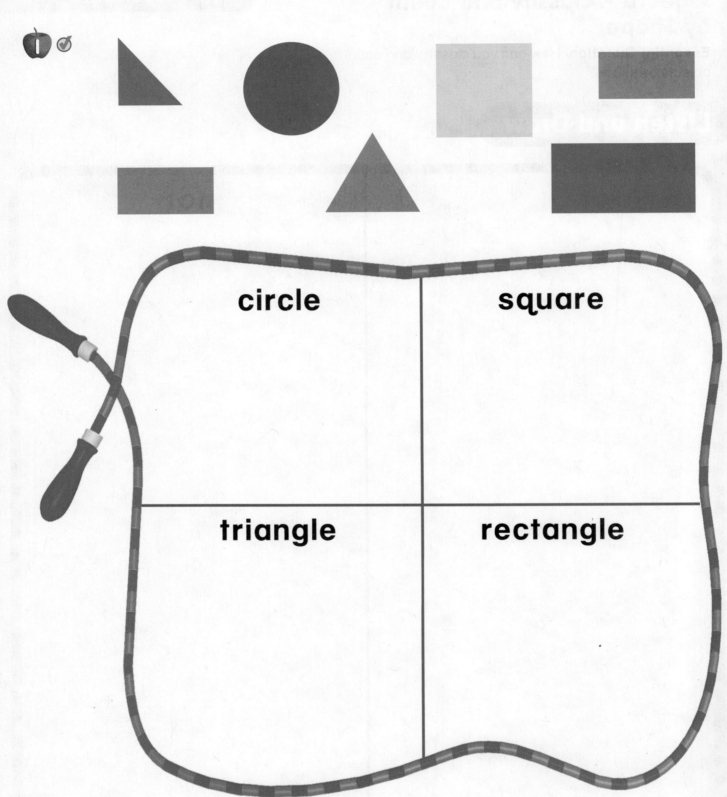

DIRECTIONS 1. Place shapes as shown. Sort and classify the shapes by the category of shape. Draw and color the shapes in each category.

②

1

circle	square
triangle	rectangle

_____
- - - - - - - - -
_____

③

2

circle	square
triangle	rectangle

_____
- - - - - - - - -
_____

④

3

circle	square
triangle	rectangle

_____
- - - - - - - - -
_____

DIRECTIONS Look at the categories in Exercise 1. Count how many in each category. **2.** Circle the categories that have one shape. Write the number. **3.** Circle the category that has two shapes. Write the number. **4.** Circle the category that has three shapes. Write the number.

PROBLEM SOLVING

❶

❷

DIRECTIONS **1.** How are these shapes sorted and classified? Draw one more shape in each category. **2.** Using the same shapes, draw to show what you know about sorting and classifying by shape in a different way.

HOME ACTIVITY • Have your child sort household objects into categories of shape.

500 five hundred

Name _____

Algebra • Classify and Count by Size

Essential Question How can you classify and count objects by size?

Listen and Draw

big	small

DIRECTIONS Sort and classify a handful of shapes by size. Draw and color the shapes.

Chapter 12 • Lesson 3

five hundred one **501**

Share and Show

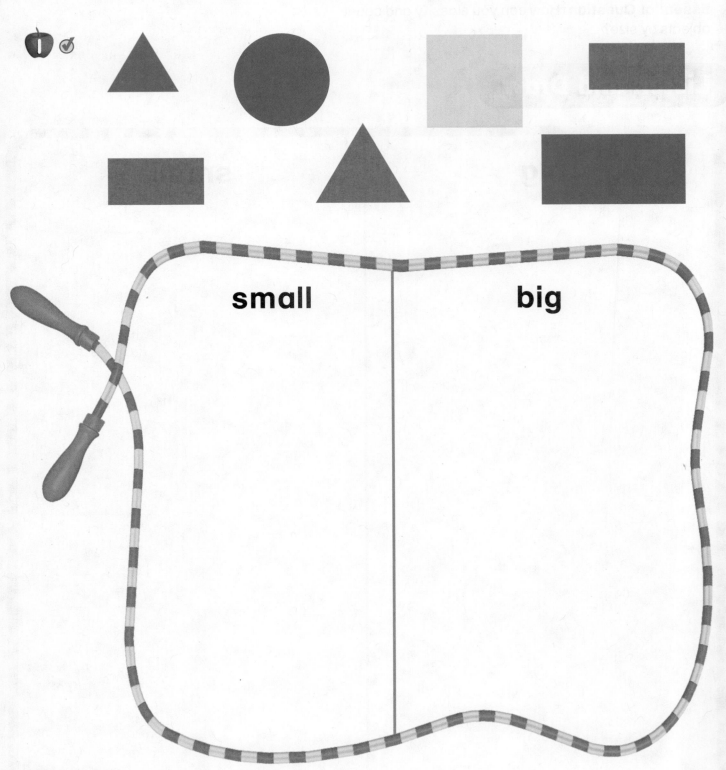

small   big

DIRECTIONS 1. Place shapes as shown. Sort and classify the shapes by the category of size. Draw and color the shapes in each category.

502 five hundred two

Name _____

3

small

big

_____

- - - - - - - - - -

_____

4

small

big

_____

- - - - - - - - - -

_____

DIRECTIONS Look at the categories in Exercise
1. Count how many in each category. **2.** Circle
the category that has three per category. Write the
number. **3.** Circle the category that has four per
category. Write the number.

HOME ACTIVITY • Have your child
sort household objects into categories
of size.

FOR MORE PRACTICE:
Standards Practice Book, pp. P239–P240

Chapter 12 • Lesson 3

Concepts and Skills

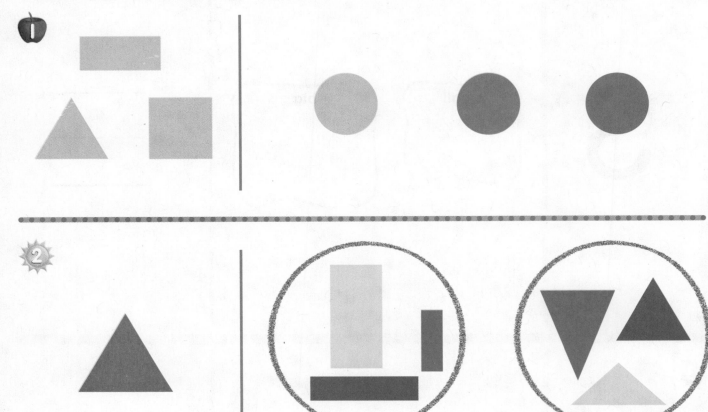

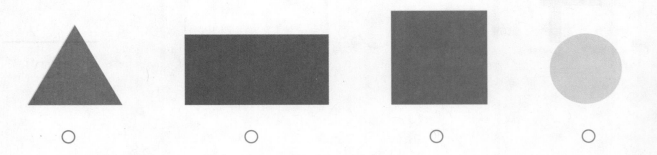

○   ○   ○   ○

DIRECTIONS **1.** Look at the set at the beginning of the row. Circle the shape that belongs in that set. **2.** Look at the shape at the beginning of the row. Mark an X on the set in which the shape belongs. **3.** Mark under the picture that shows the shape that does not belong.

Name _____

Make a Concrete Graph

Essential Question How can you make a graph to count objects that have been classified into categories?

Listen and Draw

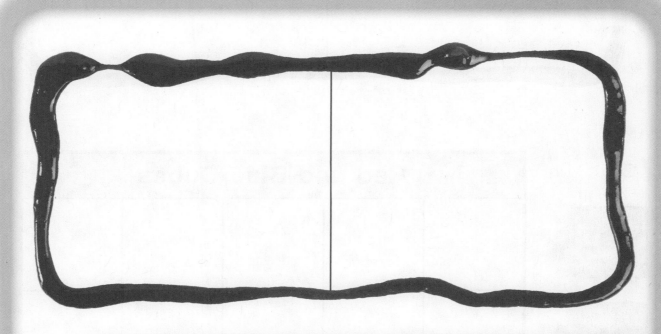

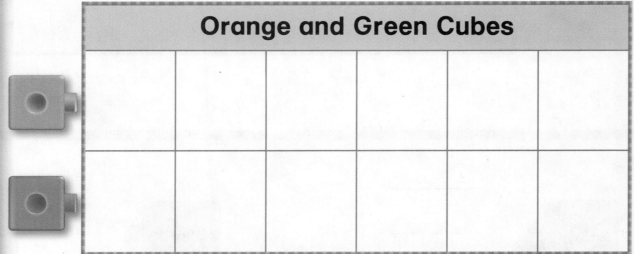

Orange and Green Cubes

DIRECTIONS Place a handful of orange and green cubes on the workspace. Sort and classify the cubes by the category of color. Move the cubes to the graph by category. Draw and color the cubes. Tell a friend how many in each category.

Chapter 12 • Lesson 4

five hundred fi...

Share and Show

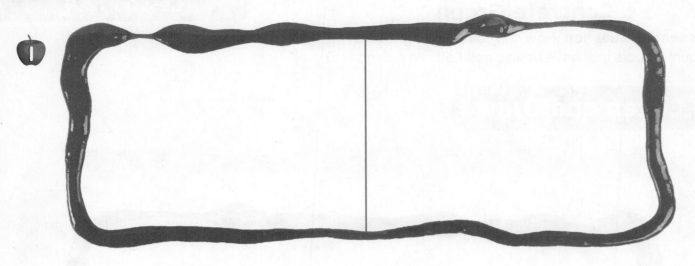

Red and Blue Cubes

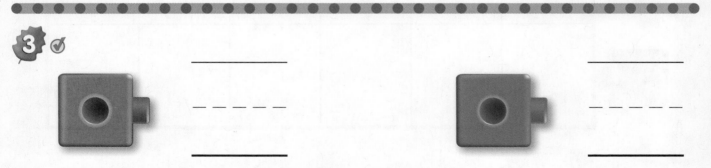

DIRECTIONS **1.** Place a handful of red and blue cubes on the workspace. Sort and classify the cubes by category. **2.** Move the cubes to the graph. Draw and color the cubes. **3.** Write how many of each cube.

five hundred six

© Houghton Mifflin Harcourt Publishing Company

Name _____

4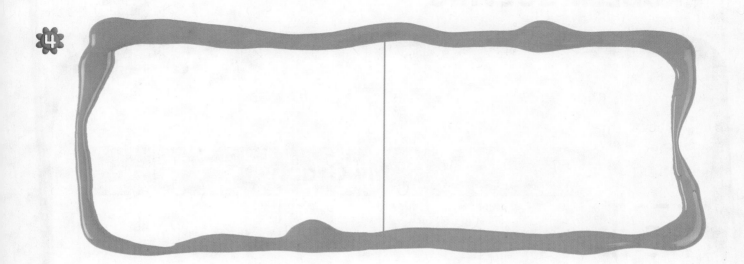

5

Green Circles and Triangles

6

_____ _____

- - - - - - - - - - - -

_____ _____

DIRECTIONS 4. Place a handful of green circles and triangles on the workspace. Sort and classify the shapes by category. **5.** Move the shapes to the graph. Draw and color the shapes. **6.** Write how many of each shape.

Chapter 12 • Lesson 4   five hundred seven **507**

PROBLEM SOLVING

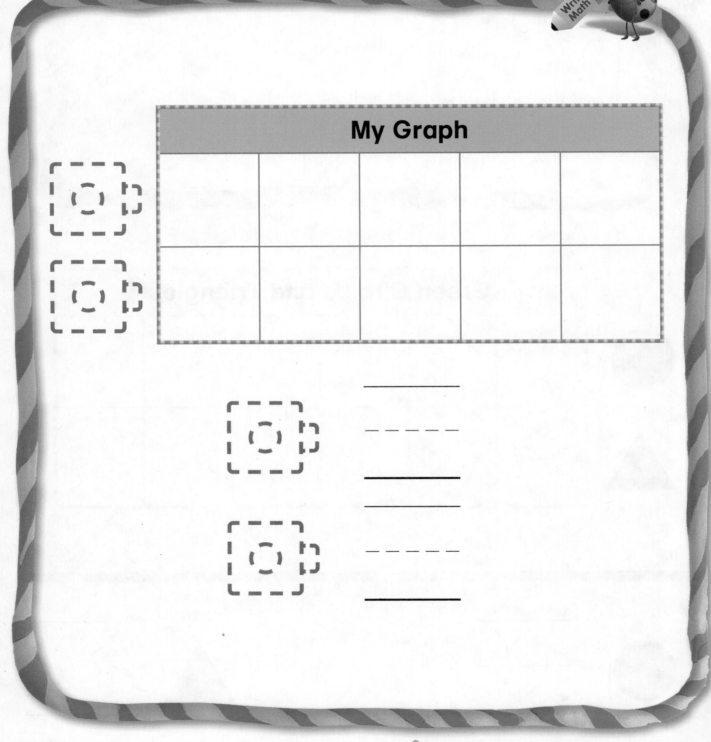

My Graph

DIRECTIONS Use five cubes of two colors. Color the cubes to show the categories. Draw and color to show what you know about making a graph with those cubes. How many in each category? Write the numbers.

HOME ACTIVITY · Have your child tell about the graph that he or she made on this page.

© Houghton Mifflin Harcourt Publishing Company

FOR MORE PRACTICE:
Standards Practice Book, pp. P241–P242

Name _____

Read a Graph

Essential Question How can you read a graph to count objects that have been classified into categories?

Listen and Draw

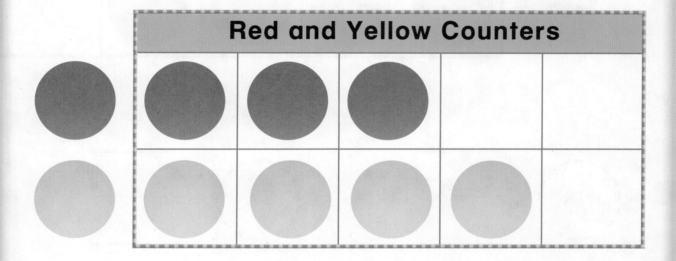

Red and Yellow Counters

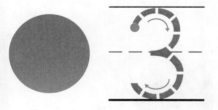

DIRECTIONS How many counters are in each category? Trace the numbers. Trace the circle to show which category has more counters.

Share and Show

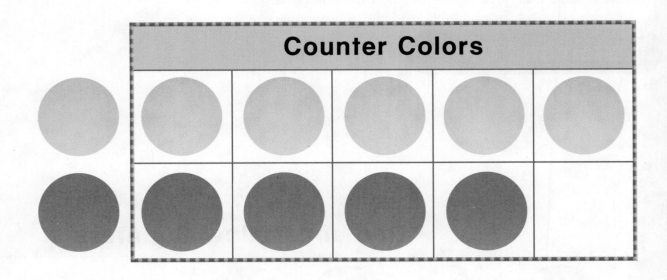

Counter Colors

 1. ( ) _____ ( ) _____

2.

DIRECTIONS 1. Color the counters to show the categories. How many counters are in each category? Write the numbers. 2. Circle the category that has more counters on the graph.

510 five hundred ten

Counter Colors

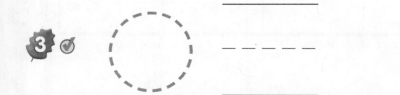

 3 ✅

⬭　———————
　　- - - - - - -
　　———————

⬭　———————
　　- - - - - - -
　　———————

4 ✅

DIRECTIONS 3. Color the counters to show the categories. How many counters are in each category? Write the numbers. **4.** Circle the category that has fewer counters on the graph.

PROBLEM SOLVING

Cube Colors

 _____

 _____

DIRECTIONS There are more blue cubes than red cubes. Draw and color to show the cubes on the graph. Count how many in each category. Write the numbers.

HOME ACTIVITY • Have your child tell about the graph he or she made on this page. Ask him or her which category has more cubes and which category has fewer cubes.

512 five hundred twelve

Name _____

Problem Solving • Sort and Count

Essential Question How can you solve problems using the strategy *use logical reasoning*?

 Unlock the Problem REAL WORLD

red   **blue**

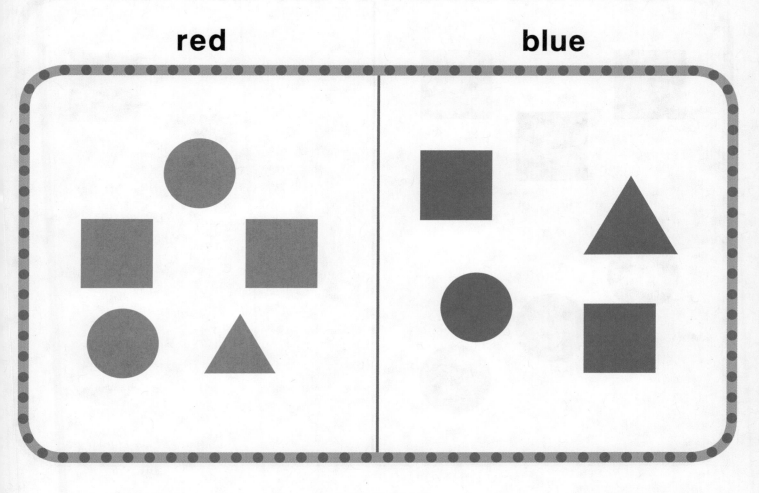

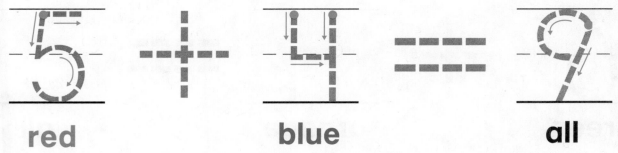

red   blue   all

DIRECTIONS Look at the sorting mat. How are the shapes sorted? How many red shapes? How many blue shapes? Add the two sets. Trace the addition sentence.

Chapter 12 • Lesson 6

Try Another Problem

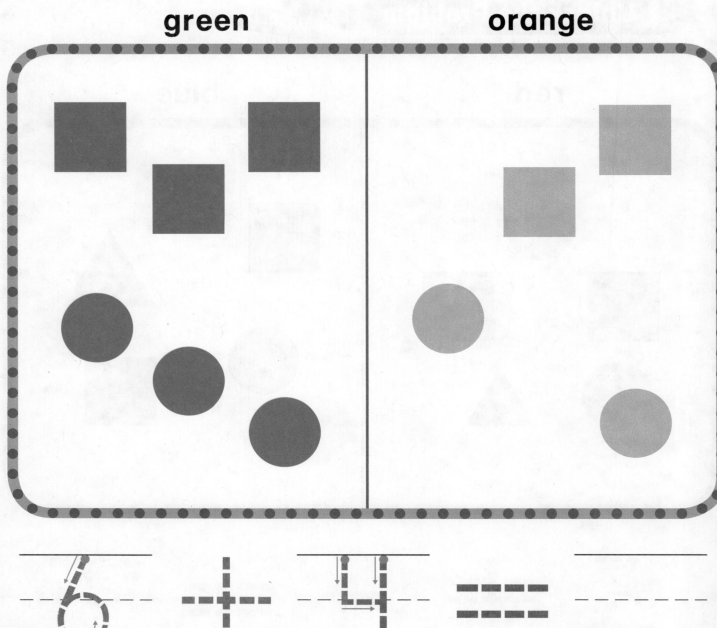

green

orange

6   +   4   =   _____

green   **orange**   **all**

DIRECTIONS 1. Look at the sorting mat. How are the shapes sorted? How many green shapes? How many orange shapes? Add the two sets. Trace and write to complete the addition sentence.

514 five hundred fourteen

Name _____

Share and Show

2 ✓

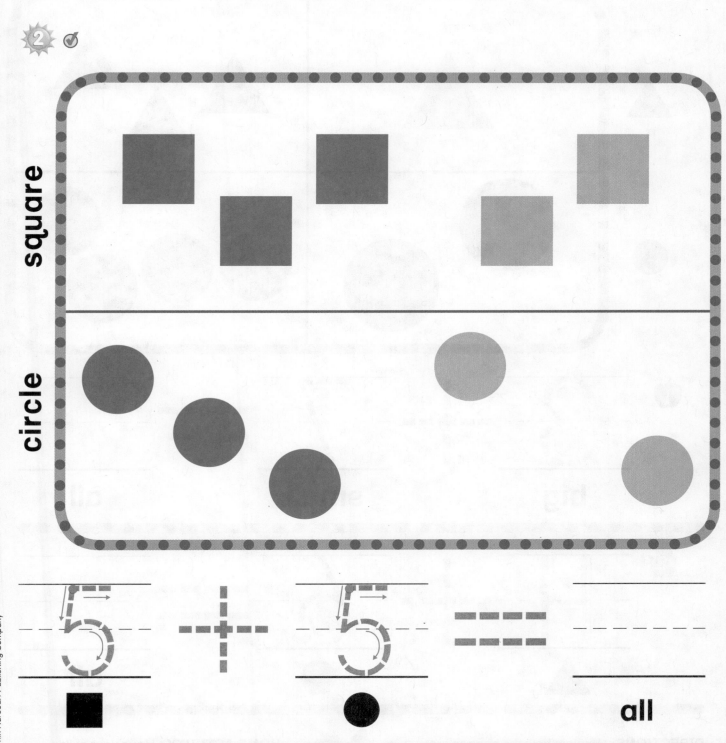

square

circle

5 + 5 = _____

■   ●   **all**

DIRECTIONS 2. Look at the sorting mat. How are the shapes sorted?
How many squares? How many circles? Add the two sets. Trace and write to
complete the addition sentence.

Chapter 12 • Lesson 6

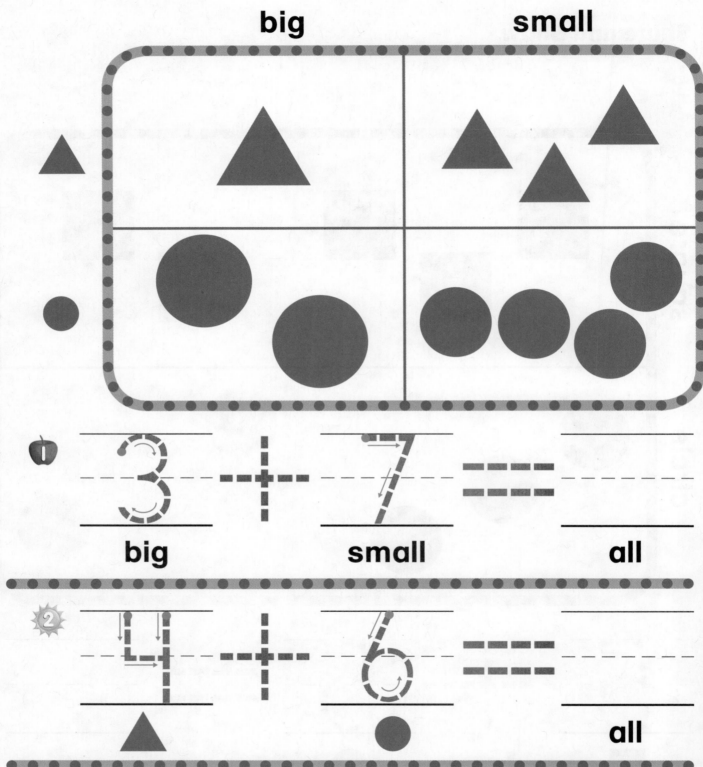

big   **small**

1. 3 + 7 = ___

big   **small**   **all**

2. 4 + 6 = ___

all

DIRECTIONS Explain how the shapes are sorted. **1.** How many big and small shapes? Trace and write to complete the addition sentence. **2.** How many triangles and circles? Trace and write to complete the addition sentence.

HOME ACTIVITY • Have your child explain the two different addition sentences on this page and tell how the shapes are sorted.

FOR MORE PRACTICE:
Standards Practice Book, pp. P245–P246

Name _____

Vocabulary

green

red

blue

Concepts and Skills

DIRECTIONS **I.** Draw a line from the color name to the color. (pp. 493–496)
2–3. Look at the shape at the beginning of the row. Circle the set of shapes
in which it belongs. **4.** Look at the size of the shapes. Mark an X on the
shape that does not belong.

5

6

7

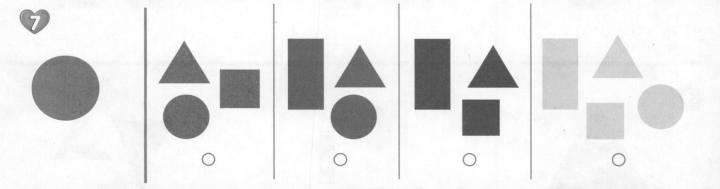

8

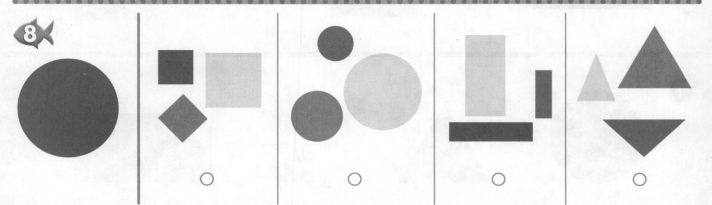

DIRECTIONS **5.** Look at the shape at the beginning of the row. Mark under the set of shapes in which it belongs. **6.** Look at the size of the shapes. Mark under the shape that does not belong. **7–8.** Look at the shape at the beginning of the row. Mark under the set of shapes in which it belongs.

Name _____

○

Red and Blue Cubes				
■	■	■	■	■
■	■	■	■	

○

Red and Blue Cubes				
■	■	■		
■	■	■	■	■

○

Red and Blue Cubes			
■	■	■	■
■	■		

○

Red and Blue Cubes			
■	■	■	
■	■	■	

10

green   blue

○ 2 + 2 = 4   ○ 5 + 2 = 7

○ 5 + 4 = 9   ○ 6 + 4 = 10

DIRECTIONS 9. Mark beside the graph that shows there are fewer red cubes than blue cubes.   **10.** Look at the sorting mat. Mark under the addition sentence that matches the sorted shapes.

Chapter 12   five hundred nineteen **519**

Performance Task

My Graph

_____

_____

- - - -

PERFORMANCE TASK This task will assess the child's understanding of sorting and graphing.

Table of Contents
Florida Lessons

Name _____

Use Nonstandard Units to Express Length

Essential Question How can you express the length of an object using nonstandard units?

Listen and Draw · REAL WORLD

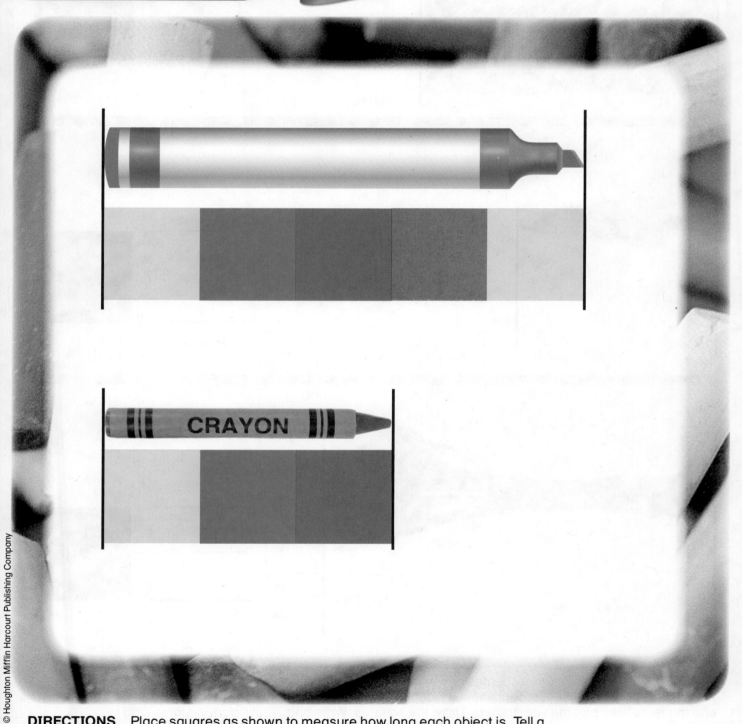

DIRECTIONS Place squares as shown to measure how long each object is. Tell a friend how many squares it took to find the length of each object.

Chapter 11 · Lesson 3A

FL1

Share and Show

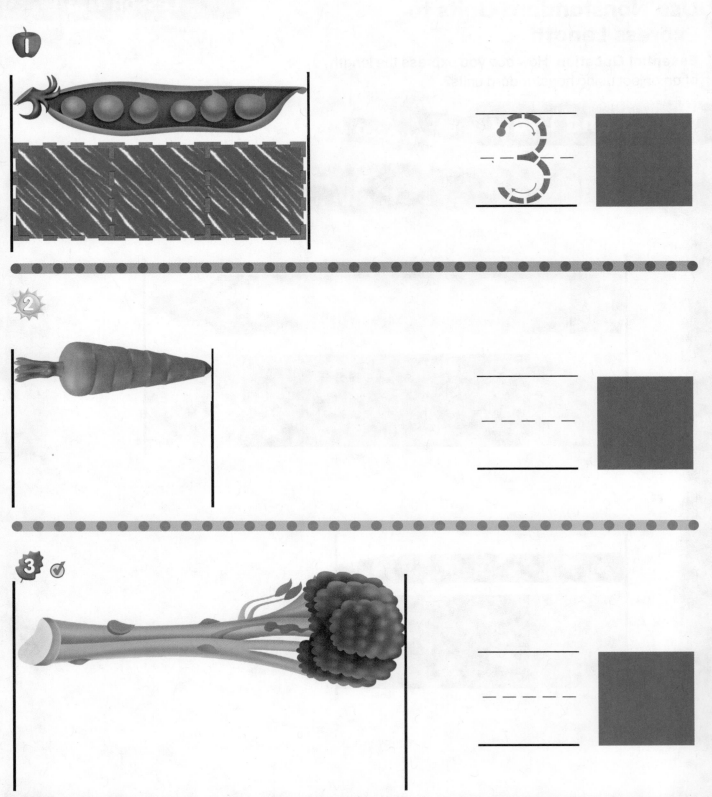

DIRECTIONS **1.** Place three squares end to end as shown to measure how long the vegetable is. Trace the squares. Trace the number that shows about how many squares long it is. **2–3.** Use squares end to end measure how long the vegetable is. Draw the squares. Write about how many squares long it is.

FL2

4

5

6

DIRECTIONS **4–6.** Use squares end to end to measure how long the vegetable is. Draw the squares. Write about how many squares long it is.

PROBLEM SOLVING

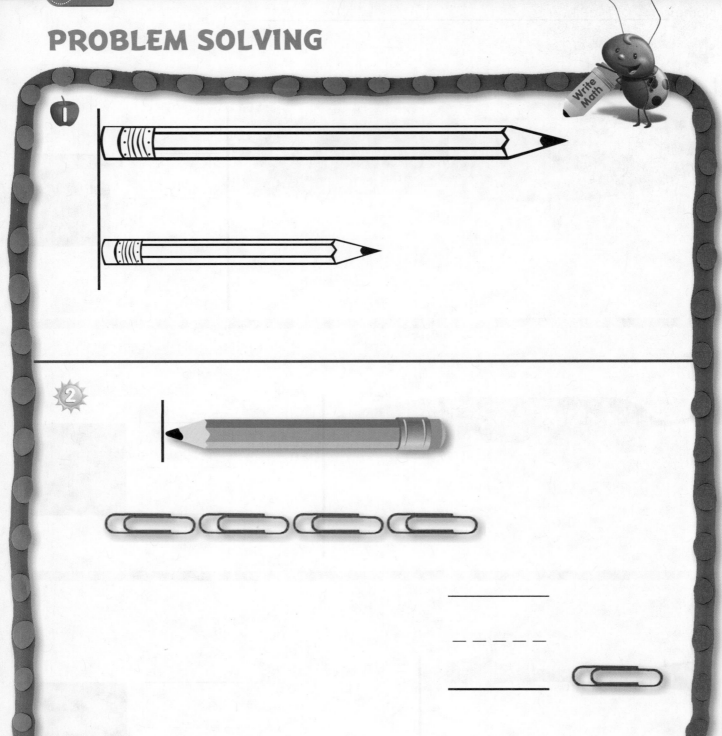

1

2

DIRECTIONS **1.** Jesse needs a pencil about 3 paper clips long. Color the pencil that Jesse needs. **2.** Katy thinks the pencil is 4 paper clips long. About how long is the pencil?

HOME ACTIVITY • Show your child a household object. Have him or her use paper clips or blocks to measure the length of the object. Then have him or her tell you how many units it took to measure the length of the object.

FOR MORE PRACTICE:
Standards Practice Book, pp. PFL1–PFL2

Picture Glossary

above [arriba]

The kite is **above** the rabbit.

add [sumar]

$$3 + 2 = 5$$

alike [igual]

and [y]

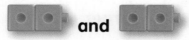

$$2 + 2$$

behind [detrás]

The box is **behind** the girl.

below [debajo]

The rabbit is **below** the kite.

beside [al lado]

The tree is **beside** the bush.

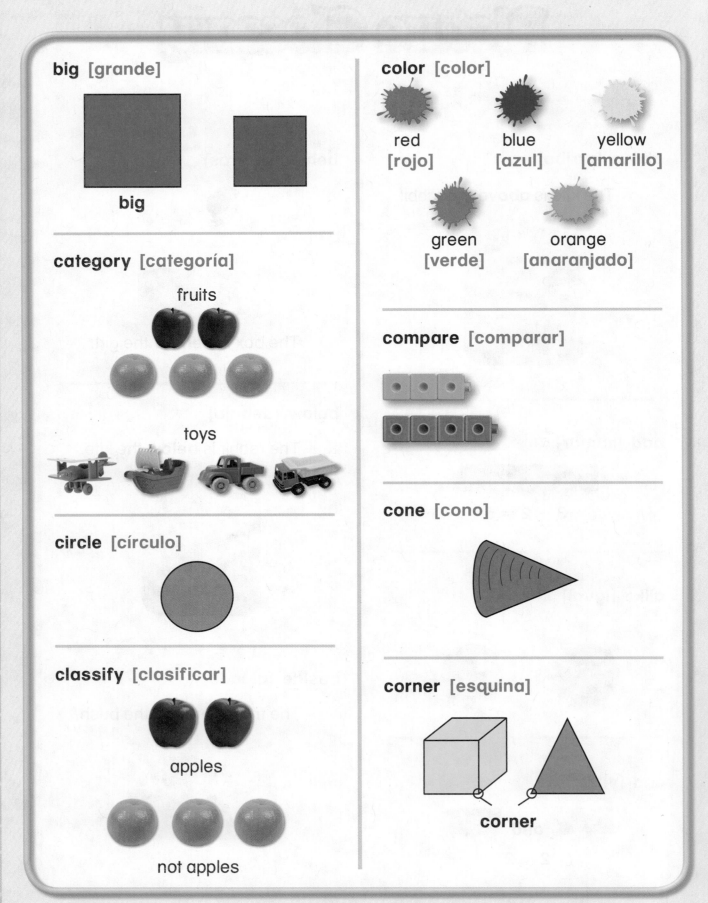

big [grande]

big

category [categoría]

fruits

toys

circle [círculo]

classify [clasificar]

apples

not apples

color [color]

red
[rojo]

blue
[azul]

yellow
[amarillo]

green
[verde]

orange
[anaranjado]

compare [comparar]

cone [cono]

corner [esquina]

corner

cube [cubo]

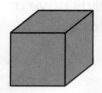

curve [curva]

curved surface
[superficie curva]

Some solids have
a **curved surface**.

cylinder [cilindro]

different [diferente]

eight [ocho]

eighteen [dieciocho]

eleven [once]

fewer [menos]

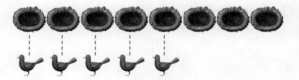

3 **fewer** birds

fifteen [quince]

fifty [cincuenta]

1	2	3	4	5	6	7	8	9	10
11	12	13	14	15	16	17	18	19	20
21	22	23	24	25	26	27	28	29	30
31	32	33	34	35	36	37	38	39	40
41	42	43	44	45	46	47	48	49	50

five [cinco]

flat [plano]

A circle is a **flat** shape.

flat surface [superficie plana]

Some solids have a **flat surface**.

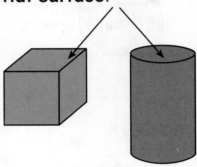

four [cuatro]

fourteen [catorce]

graph [gráfica]

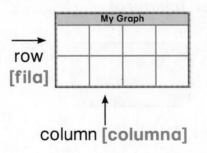

row [fila]

column [columna]

greater [mayor]

9 is **greater** than 6

6

9

heavier [más pesado]

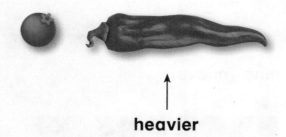

heavier

hexagon [hexágono]

in front of [delante de]

The box is **in front of** the girl.

is equal to [es igual a]

3 + 2 = 5

3 + 2 **is equal to** 5

larger [más grande]

2   3

A quantity of 3 is **larger** than a quantity of 2.

less [menor/menos]

9 is **less** than 11

9

11

lighter [más liviano]

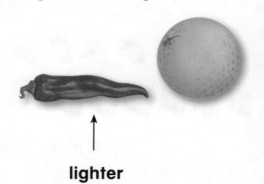

lighter

longer [más largo]

 **longer**

match [emparejar]

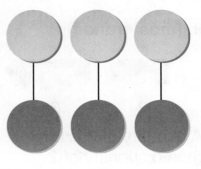

minus − [menos]

4 − 3 = 1

4 **minus** 3 is equal to 1

more [más]

2 **more** leaves

next to [al lado de]

The bush is **next to** the tree.

nine [nueve]

nineteen [diecinueve]

one [uno]

one hundred [cien]

1	2	3	4	5	6	7	8	9	10
11	12	13	14	15	16	17	18	19	20
21	22	23	24	25	26	27	28	29	30
31	32	33	34	35	36	37	38	39	40
41	42	43	44	45	46	47	48	49	50
51	52	53	54	55	56	57	58	59	60
61	62	63	64	65	66	67	68	69	70
71	72	73	74	75	76	77	78	79	80
81	82	83	84	85	86	87	88	89	90
91	92	93	94	95	96	97	98	99	100

ones [unidades]

3 ones

pairs [pares]

3

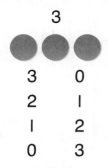

3	0
2	1
1	2
0	3

number **pairs** for 3

plus + [más]

2 **plus** 1 is equal to 3

2 + 1 = 3

rectangle [rectángulo]

roll [rodar]

same height
[de la misma altura]

same length [del mismo largo]

same number
[el mismo número]

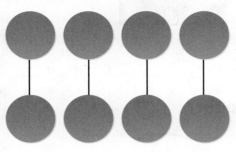

same weight [del mismo peso]

seven [siete]

seventeen [diecisiete]

shape [forma]

shorter [más corto]

shorter

side [lado]

side

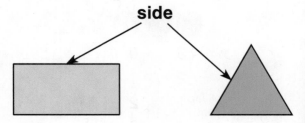

sides of equal length [lados del mismo largo]

six [seis]

sixteen [dieciséis]

size [tamaño]

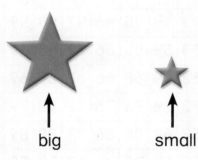

big   small

slide [deslizar]

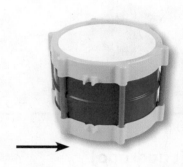

small [pequeño]

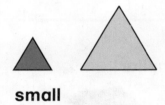

small

solid [sólido]

solid

A cylinder is a **solid** shape.

sphere [esfera]

square [cuadrado]

stack [apilar]

subtract [restar]

Subtract to find out how many are left.

taller [más alto]

taller

ten [diez]

tens [decenas]

1	2	3	4	5	6	7	8	9	10
11	12	13	14	15	16	17	18	19	20
21	22	23	24	25	26	27	28	29	30
31	32	33	34	35	36	37	38	39	40
41	42	43	44	45	46	47	48	49	50
51	52	53	54	55	56	57	58	59	60
61	62	63	64	65	66	67	68	69	70
71	72	73	74	75	76	77	78	79	80
81	82	83	84	85	86	87	88	89	90
91	92	93	94	95	96	97	98	99	100

tens

thirteen [trece]

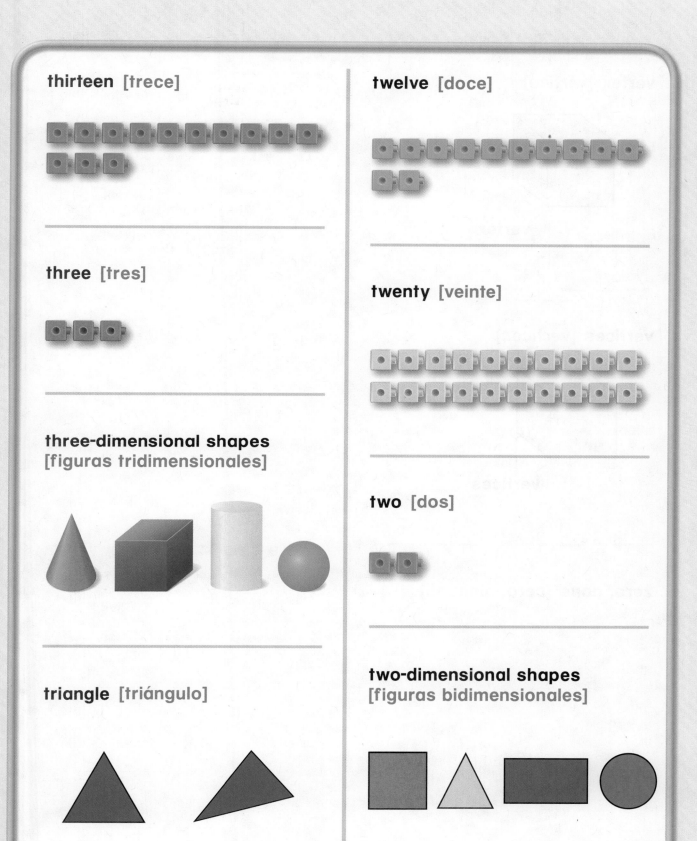

twelve [doce]

three [tres]

twenty [veinte]

three-dimensional shapes
[figuras tridimensionales]

two [dos]

triangle [triángulo]

two-dimensional shapes
[figuras bidimensionales]

vertex [vértice]

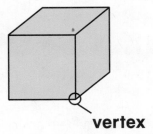

vertex

vertices [vértices]

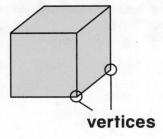

vertices

zero, none [cero, ninguno]

zero fish

Photo Credits

KEY: (t) top, (b) bottom, (l) left, (r) right, (c) center, (bg) background, (fg) foreground, (i) inset

Cover: (bg) Siede Preis; (br) PhotoDisc/Getty Images; (bg) CreativeAct-Seasonal Series/Alamy; (bg) Alamy; (bg) Tom Brakefield/Stockbyte/Getty Images; (bg) Corbis.

Copyright Page: (tl) Danita Delimont/Alamy Images; (blind) (br) Don Hammond/Design Pics/Corbis.

Author Page: altrendo nature/Getty Images; (bg) (2 images) Purestock/Getty Images; (br) Daniel J. Cox/Getty Images.

Table of Contents: v (bg) Wally Eberhart/PictureArts/Corbis; vi (bl) PhotoDisc/Getty Images; (bg) Corel Stock Library Royalty Free; vii (bg) Herbert Kehrer/Corbis; viii (bl) altrendo nature/Getty Images; ix (bg) Emma Lee/Life File/Getty Images; x (l) LHB Photo/Alamy; xi (bc) PhotoDisc/Getty Images; (bg) Corel Stock Photography; xii (bg) Andrea Rugg Photography/Corbis.

Critical Area: 9 (bg) Wally Eberhart/PictureArts/Corbis; 19 (2 images) Artville/Getty Images; 28 (bl) PhotoDisk/Getty Images; 29 (5 images) Ramesh Racha/Alamy; 37 (bg) Corbis Premium RF/Alamy; 41 (bg) PhotoDisk/Getty Images; 43 (c) (2 images) PhotoDisc/Getty Images; 55 (t) Ingram Publishing/Alamy; (bl) PhotoDisc/Getty Images; 57 (bg) Connie Coleman/Getty Images; 61 (bg) Steffen Hauser/Botanikfoto/Alamy; 65 (bg) Corbis; 69 (bg) Corel Stock Photo Library; 85 (bg) Dennis MacDonald/Alamy; 101 (cl) PhotoDisk/Getty Images; (cl) (7 images) PhotoDisk/Getty Images; 103 (tr) (5 images) Stockbyte/Getty Images; (br) (7 images) PhotoDisk/Getty Images; 104 (cr) (7 images) Corbis; 105 (bg) Jim Mires/Alamy; 113 (bg) PhotoDisk/Getty Images; 117 (bc) (9 images) Corbis; 120 (t) (20 images) Ann Triling/Shutterstock; 121 (bg) Dennis Palmer/Alamy; 126 (cr) (28 images) PhotoDisc/Getty Images; 129 (bg) Herbert Kehrer/Corbis; 133 (bg) PhotoDisk/Getty Images; 145 (bg) PhotoDisk/Getty Images; 149 (bg) PhotoDisk; 153 (tc) (6 images) Stockbyte/Getty Images; (8 images) (bc) Corbis; (bg) Image Source Pink/Alamy; 154 (tl) (8 images) PhotoDisc/Getty Images; (tr) (10 images) Andrew Paterson/Alamy; (cr) (6 images) Corbis; (bl) (6 images) Comstock/Getty Images; (br) (9 images) Artville/Getty Images; 155 (tl) (17 images) Stockbyte/Getty Images; (tr) (7 images) Eyewire/Getty Images; 156 (tl) (8 images) Comstock/Getty Images; (tr) (8 images) PhotoDisk/Getty Images; 157 (bg) PhotoDisk/Getty Images; 162 (c) (8 images) Stockdisk/Getty Images; (c) (7 images) Stockbyte/Getty Images; 163 (c) (7 images) PhotoDisk/Getty Images; (c) (10 images) Corbis; 165 (bg) Pernilla Bergdahl/GAP Photos/Getty Images; 173 (border) Artville/Getty Images; 181 (bg) PhotoDisk/Getty Images; 185 (border) Corbis; 197 (border) PhotoDisk/Getty Images; 201 (border) PhotoDisk/Getty Images; 205 (border) Artville/Getty Images; 209 (border) Comstock/Getty Images; 213 (border) PhotoDisk/Getty Images; 221 (bg) Randy Wells/Corbis; 229 (border) Corbis; 237 (bg) Corbis; 257 (bg) Emma Lee/Life File/Getty Images; 258 (cl) (10 images) Stockbyte/Getty Images; (cr) (9 images) PhotoAlto/Getty Images; (bl) (8 images) Artville/Getty Images; (br) (10 images) Ramesh Racha/Alamy; 261 (bg) PhotoDisk/Getty Images; 262 (tc) (11 images) Ingram; 263 (tc) (12 images) Ingram; 269 (border) Siede Preis/PhotoDisk/Getty Images; 270 (cr) (13 images) Ingram; (bc) Artville/Getty Images; 271 (tc) (13 images) Ingram; 277 (bg) John Glover/Getty Images; 278 (tc) (15 images) Ingram; 281 (border) PhotoDisk/Getty Images; 284 (cl) (26 images) Ingram; 285 (border) Artville/Getty Images; 286 (tc) (16 images) Stockbyte/Getty Images; 287 (tc) (17 images) Artville/Getty Images; 293 (border) Siede Preis/PhotoDisk/Getty Images; 294 (tc) (19 images) Artville/Getty Images; 295 (tc) (19 images) Artville/Getty Images; 302 (cl) (19 images) Stockbyte/Getty Images; 305 (bg) Nicholas Eveleigh/Stone/Getty Images; 311 (tc) (20 images) Ramesh Racha/Alamy; (c) (19 images) Artville/Getty Images; 314 (tc) (29 images) Ingram; (bc) (18 images) Artville/Getty Images; 315 (tr) (19 images) Artville/Getty Images; 316 (t) (11 images) Ingram; 317 (bg) Lew Robertson/Getty Images; 321 (bg) PhotoDisk/Getty Images; 324 (tc) (34 images) Artville/Getty Images; (bl) (20 images) Ingram; (c) (31 images) Brian Hagiwara/Getty Images; 325 (border) Artville/Getty Images; 337 (border) Artville/Getty Images; 341 (tc) (20 images) Ingram; 343 (tc) (86 images) Ingram.

Critical Area: 351 Comstock Images/Jupiter Images/Getty Images; 353 (bg) LHB Photo/Alamy; 454 (bg) Purestock/Alamy; 455 (bg) PHOTO 24/Getty Images; 456 (bg) John Ferro Sims/Alamy; 457 (bg) Juniors Bildarchiv/Alamy Images; 458 (bg) Roderick Chen/All Canada Photos/Corbis; 461 (bg) Andrea Rugg Photography/Corbis; 465 (bg) Corbis Premium RF/Alamy; 469 (bg) Garry Gay/Alamy; 473 Bryan Mullennix/Getty Images; 478 (cl) PhotoDisk; (bl) Artville/Getty Images; 481 Corbis Premium RF/Alamy; 483 Stockbyte/Getty Images; 485 Stockbyte/Getty Images; 487 (cl) PhotoDisk; (cr) Artville/Getty Images; NiKreationS/Alamy; Stockbyte/Getty Images; 489 (bg) PhotoAlto/Laurence Mouton/Getty Images; FL1 Corbis Premium RF/Alamy.

Picture Glossary: 522 Artville/Getty Images; D. Hurst/Alamy; C Squared Studios/PhotoDisc/Getty Images.

All other photos Houghton Mifflin Harcourt libraries and photographers; Guy Jarvis, Weronica Ankarorn, Eric Camden, Don Couch, Doug Dukane, Ken Kinzie, April Riehm, and Steve Williams.